Mathematics for Engineers

Scrivener Publishing
100 Cummings Center, Suite 541J
Beverly, MA 01915-6106

Publishers at Scrivener
Martin Scrivener (martin@scrivenerpublishing.com)
Phillip Carmical (pcarmical@scrivenerpublishing.com)

Mathematics for Engineers

Ritu Shrivastava
Ramakant Bhardwaj
and
Satyendra Narayan

Scrivener Publishing

WILEY

This edition first published 2025 by John Wiley & Sons, Inc., 111 River Street, Hoboken, NJ 07030, USA and Scrivener Publishing LLC, 100 Cummings Center, Suite 541J, Beverly, MA 01915, USA

For more information about Scrivener publications please visit www.scrivenerpublishing.com.

Wiley Global Headquarters
111 River Street, Hoboken, NJ 07030, USA

For details of our global editorial offices, customer services, and more information about Wiley products visit us at www.wiley.com.

Library of Congress Cataloging-in-Publication Data

ISBN 9781394214587

Front cover images supplied by Adobe Firefly
Cover design by Russell Richardson

Set in size of 11pt and Minion Pro by Manila Typesetting Company, Makati, Philippines

Printed in the USA

10 9 8 7 6 5 4 3 2 1

Contents

Preface

Mathematics without practice is incomplete. Practice of mathematics without understanding theory is meaningless. Theory and practice are two sides of the same coin. Like the two sides of a coin, theory and practice have their differences, and they are inseparable. *Mathematics for Engineers* is designed to explain theory and practice together in a meaningful way, providing engineers with a comprehensive understanding of fundamental mathematical concepts crucial for their academic and professional endeavors. The depth and breadth of this textbook offers comprehensive coverage of all the basic mathematical tools needed for first-year engineering students, as well as a reference for veteran engineers and industry professionals.

This textbook is originated from a one-semester course dedicated to introductory engineering mathematics taught in many universities and engineering colleges over the past several years primarily to first-year engineering students. This covers a wide range of syllabus requirements globally. In particular, the textbook is highly suitable for the latest National Certificate and Diploma courses and Vocational Certificate of Education in science & engineering programs. First year undergraduates who need some remedial mathematics may also find it helpful in meeting their needs. Overall, the textbook provides a solid foundation in mathematical principles, which may enable students to solve mathematical, scientific, and associated engineering principles. Key Features of this book are highlighted below.

Chapter 1: Fractions, Decimals, and Percentages lays the groundwork by revisiting fundamental arithmetic concepts, such as fractions, decimals, and percentages. From understanding the basics of fractions to performing algebraic operations and mental math tricks, it equips leaners with essential skills for more complex mathematical endeavors.

Chapter 2: Ratios and Proportions exploring ratios, rates, and proportions delves into the principles governing the relative magnitudes of quantities. They play pivotal roles in engineering, influencing everything from structural design to financial analysis. Learners will learn techniques for simplifying ratios, finding equivalent ratios, and solving problems involving direct and inverse proportions.

Chapter 3 and 4: 2D & 3D Geometry are fundamental to engineering, and it covers essential concepts related to 2D and 3D shapes, perimeter, area, surface area, and volume calculations. Understanding geometric principles is important, crucial, and required for designing structures and analyzing spatial relationships in engineering applications.

Chapter 5: Algebra and Graphs provides algebraic skills that are indispensable for engineers. It covers equations, expressions, and graphical representations of mathematical relationships. From linear to quadratic equations and beyond, learners will explore various algebraic techniques required for problem-solving in engineering contexts.

Chapter 6: Exponentials & Logarithms are the powerful mathematical tools. They are used in a wide range of applications in engineering problems. It elucidates the rules governing these concepts, their graphical representations, and their role in modeling growth and decay phenomena.

Chapter 7: Trigonometry is indispensable for analyzing angles, distances, and relationships within geometric structures. It covers trigonometric ratios, Pythagoras' theorem, trigonometric equations, and applications such as finding angles of elevation and depression.

Chapter 8 and 9: Differential Calculus & Integration provide engineers with powerful tools for analyzing rates of change, optimization problems, and modeling dynamic systems. Differentiation, integration, and their applications in engineering contexts are covered completely.

Chapter 10: Probability plays an important and crucial role in engineering decision-making problems and risk assessment. This chapter introduces principles of counting, permutations, combinations, and various techniques for calculating probabilities of single and compound events.

Each chapter is dedicated to building a strong foundation, engineering applications, problem-solving skills, culturally relevant context, and practical problem-based learning (PBL) exercises and experiments to reinforce learner's understanding and application of mathematical concepts in real-world engineering scenarios.

The authors hope this textbook serves as a valuable resource in the student's journey to mastering mathematics for engineering and as a reference for the veteran engineer or industry professional. It does not matter whether you are just embarking on your academic studies or a practicing engineer seeking to enhance your mathematical skills and proficiency. This book is designed to support each individual's educational and professional development.

The authors of this book would like to acknowledge the use of freely available mathematical figures and images sourced from various online repositories. These resources have been invaluable in illustrating key concepts and enhancing the visual appeal of this book.

This book builds upon the foundational work of many brilliant mathematicians. Their groundbreaking discoveries have shaped our understanding of the world and continue to inspire generations of researchers.

1

Fractions, Decimals, and Percentages

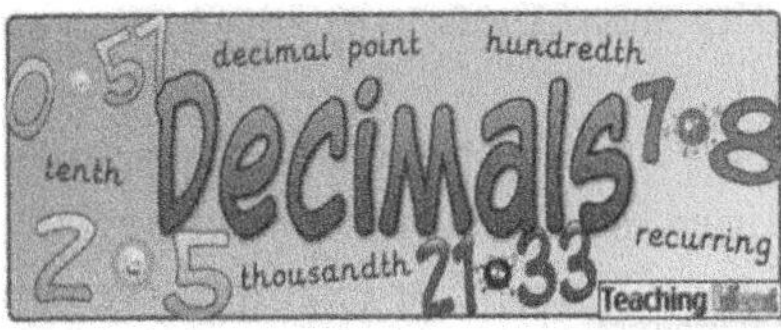

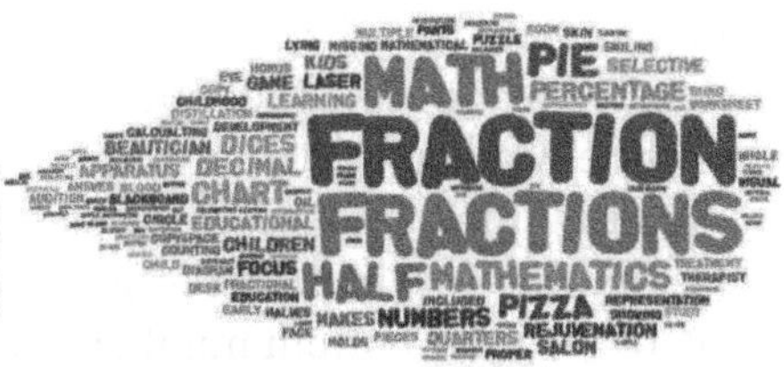

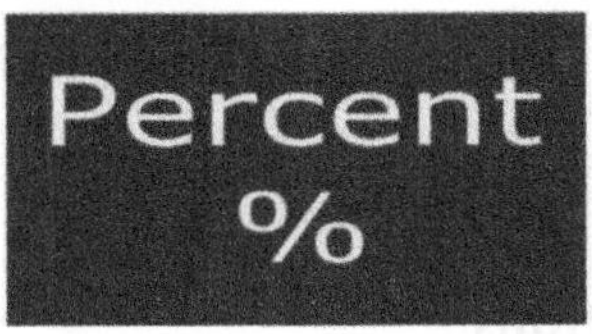

Objectives: This chapter
Introduces terminology associated with Fractions, decimals, and percentages.
Explains the methods to convert Fractions, decimals, and percentages.

1.1 Introduction

You may frequently have seen discounts written as 50% off or $\frac{1}{2}$ rate but never as 0.5 times the original rate in malls and stores. Also, when we convert currency, the exchange rates never show us as a decimal or a percentage.

These presentations are mathematically equivalent and are different ways of saying the same results.

The percentage considered is out of 100. A fraction is considered a number less than a whole. A Decimal is a number where the 'whole' part of a number and the 'fractional' part is separated by a decimal point.

Ritu Shrivastava, Ramakant Bhardwaj and Satyendra Narayan. Mathematics for Engineers, (1–36)

1.2 Fractions

1.2.1 Definition of Fractions

A fraction is simply a value that represents part of a whole. The critical point that one must understand is that when we speak about a fraction, we generally refer to some whole thing. Whenever we divide a whole into two parts, both must be equal. We will always speak of some *whole* amount divided into a specified number of equally sized pieces. In short, we can think briefly about a fraction as given below.

Notation for a fraction is the slash (/) written between the two numbers. Another way of representing fractions is to keep a top number, the numerator, and a bottom number, the denominator. For example, we can write a fraction like $\frac{3}{5}$. "3," the top number, is the numerator, and "5," the bottom, is the denominator.

1.2.2 Types of Fractions

We can see many different fractions, but the most popular are three types of fractions in Mathematics.

Common Fraction or Proper Fraction

When a fraction representing a part of the whole is known as a proper fraction (For example: in the case of a proper fraction, the denominator is always greater than the fraction's numerator).

Look at the fraction $\frac{3}{8}$. Can you tell if this is a proper fraction or not?

This fraction is a proper fraction as the denominator '8' is greater than the numerator '3'.

$\frac{8}{3}$ is not a proper fraction as the numerator '8' is greater than the denominator '3'.

Improper Fraction

A fraction that is not a proper fraction is called an improper fraction (e.g., the denominator of the improper fraction is less than or equal to its numerator).

$\frac{5}{3}$ is an improper fraction as the numerator '5' is greater than the denominator '3'.

$\frac{3}{5}$ is not an improper fraction as the numerator '3' is less than the denominator '5'.

Mixed Fraction

Those fractions expressed as a sum of a whole and a proper fraction number are called mixed fractions.

$\frac{9}{4} = 2\frac{1}{4}$ is a mixed fraction, and it can also be written as $2 + \frac{1}{4}$.

***How to Convert an Improper Fraction to a Mixed Fraction?**

Divide the numerator until the remainder becomes less than the denominator, then the mixed fraction will be in the format shown below.

Integer part (Reminder/Denominator)

- **Step 1:** Divide the given fraction's numerator by its denominator and note its quotient and remainder.
- **Step 2:** The integer part you got by division will be the integer part for a mixed fraction.
- **Step 3:** Leave the denominator the same as the original.
- **Step 4:** Write the improper fraction into the mixed fraction.

For example: Convert $\frac{17}{8}$ into a mixed fraction.

- **Step 1:** $\frac{17}{8}$ has '2' as the quotient and '1' as the remainder.
- **Step 2:** In this case, '2' is the integer.
- **Step 3:** Here, the denominator is '8'.
- **Step 4:** Hence, $\frac{17}{8}$ is changed to a mixed fraction $2\frac{1}{8}$.

***How to Convert a Mixed Fraction to an Improper Fraction?**

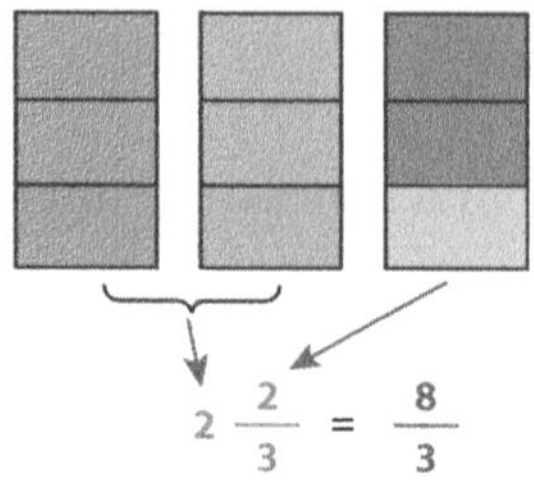

- **Step 1:** Multiply the denominator of the given fraction with the integer.
- **Step 2:** Add the fraction's numerator to the result in step 1.
- **Step 3:** Leave the denominator the same as the original.
- **Step 4:** Write the mixed fraction as an improper fraction. Reduce the result to the simplest form.

For example: Convert $2\frac{1}{8}$ into Improper fractions.

- **Step 1:** Multiply 8 with 2 in, $2\frac{1}{8}$, $8 \times 2 = 16$
- **Step 2:** Result in step 1 = 16 and numerator = 1, 1+ 16 =17.
- **Step 3:** Here, the denominator is '8'.
- **Step 4:** Hence, the improper fraction obtained is $\frac{17}{8}$ which is already in its simplest form.

Complex Fractions

A fraction in which the numerator, denominator, or both are fractions is called a complex fraction. $\frac{3/5}{7}, \frac{5/7}{11}$ are examples of complex fractions.

Equivalent Fraction

Fractions that have the same final value are known as Equivalent Fractions. For example:

$\frac{4}{8}$	=	$\frac{2}{4}$	=	$\frac{1}{2}$
(Four-Eighths)		**(Two-Fourths)**		**(One-Half)**
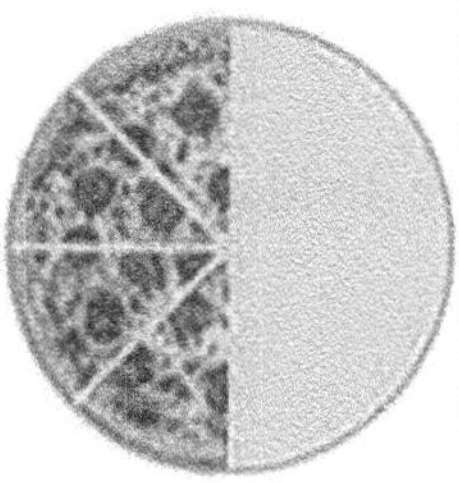	=	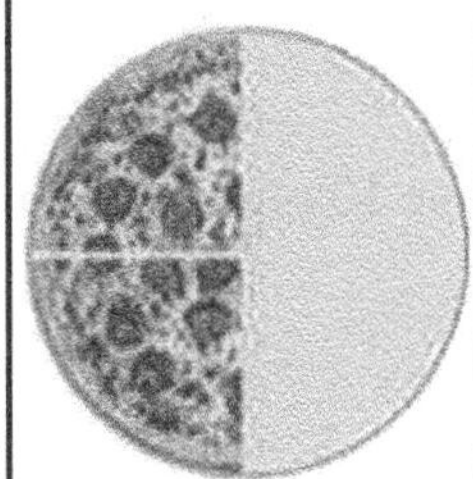	=	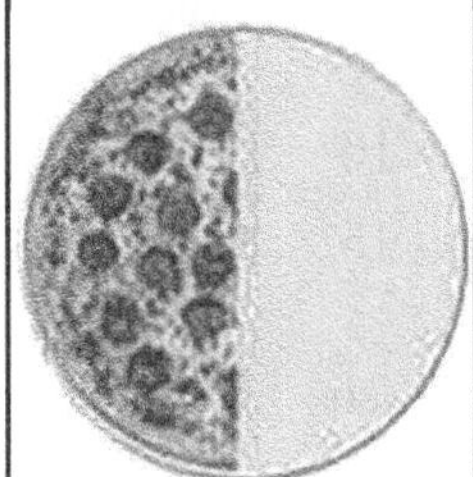

For any problem, the final answer should be written in the simplest form.

Like Fractions

Like fractions are fractions that have the same denominator. $\frac{7}{15}, \frac{11}{15}$ etc., are like fractions.

Unlike Fractions

Unlike fractions are fractions that have different denominators. $\frac{2}{15}, \frac{5}{13}, \frac{7}{17}$ etc., are unlike fractions.

Unit Fractions

A fraction with '1' as the numerator and a "positive integer" as the denominator is a unit fraction. $\frac{1}{1}, \frac{1}{2}, \frac{1}{3}, \frac{1}{4}$ are called unit fractions.

1.2.3 Simplest Form or Fraction in Lowest Term

A fraction in simplest forms is just a ratio, $\frac{p}{q}$, where p and q share no common prime factors.

- **Step 1:** Identify a number that will divide into both the numerator and the denominator.
- **Step 2:** Divide the numerator and denominator by this number.

For example:

$$\frac{10}{15} = \frac{10/5}{15/5} = \frac{2}{3} \text{(because 10 and 15 can both be divided by 5)}$$

Also,

$$\frac{4}{8} = \frac{4/4}{8/4} = \frac{1}{2} \text{(because 4 and 8 can both be divided by 4)}$$

If a fraction's numerator and denominator share only '1' as a common factor, it is said to be in its simplest form.

EXAMPLE 1: Convert $\frac{225}{180}$ to its lowest form.

SOLUTION: The factorization method determines the 225 and 180's highest common factor (HCF).

The factors of 225 are 1, 3, 5, 9, 15, 25, 45, 75, and 225.

The factors of 180 are 1, 2, 3, 4, 5, 6, 9, 10, 12, 15, 18, 20, 30, 36, 45, 60, 90 and 180.

The common factors of 225 and 180 are 1, 3, 5, 9, 15, and 45.

The highest common factor (HCF) of 225 and 180 is 45.

$$\frac{225}{180} = \frac{225/45}{180/45} = \frac{5}{4} \text{(In the simplest form)}$$

1.2.4 Comparing Fractions

Looking at two given fractions and then figuring out which one is greater is known as Comparing fractions.

Comparing Like Fractions
It is always easy to compare two or more like fractions because their denominators are the same. Compare only the numerators of the given like fractions to identify which is greater. If there are two like fractions, the fraction with a greater numerator is greater.

For example:

$$\frac{9}{13} > \frac{5}{13}$$

To arrange more than two like fractions in ascending or descending order, we must arrange their numerators in ascending or descending order, respectively.

Comparing Unlike Fractions
If denominators are not the same, follow these steps:

- **Step 1:** Find the LCM (Least Common Multiple) of their denominators for the given fractions.
- **Step 2:** Reduce each fraction to its equivalent fraction with a denominator like the LCM obtained in step 1.
- **Step 3:** Finally, arrange the fractions in ascending or descending order by arranging the numerator in ascending or descending order.

EXAMPLE 2: Sarah has three-fourths of a pizza, and Aisha has two-thirds. If both pizzas are the same size, which girl has more pizza?

SOLUTION: Fractions $\frac{3}{4}$ and $\frac{2}{3}$ have, unlike denominators, unlike numerators. We must change these fractions into equivalent fractions with a common denominator to simplify comparisons.

$$\text{Sarah: } \frac{3}{4} = \frac{3\times 3}{4\times 3} = \frac{9}{12}$$

$$\text{Aisha: } \frac{2}{3} = \frac{2\times 4}{3\times 4} = \frac{8}{12}$$

$$\frac{3}{4} > \frac{2}{3} \text{ since } \frac{9}{12} > \frac{8}{12}$$

Since nine-twelfths is greater than eight-twelfths, three-fourths is greater than two-thirds. Therefore, Sarah had more pizza.

1.2.5 Algebra of Fractions

Addition & Subtraction of Fractions

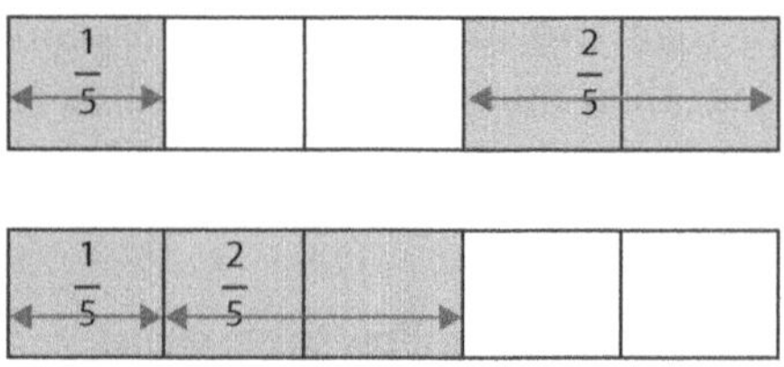

$$\frac{1}{5} + \frac{2}{5} = \frac{3}{5}$$

***How to Add and Subtract Fractions?**

- **Step 1:** Convert fractions to equivalent fractions with a common denominator.
- **Step 2:** Add or subtract the numerators only.
- **Step 3:** Leave the denominator the same.
- **Step 4:** Change the answer to the lowest terms

EXAMPLE 3: Add $\frac{1}{2}$ and $\frac{7}{8}$

SOLUTION: $\frac{1}{2}+\frac{7}{8}$ Here, the common denominator is "8" because both 2 and 8 are factors of 8

$$\frac{1}{2}=\frac{4}{8} \text{ and } \frac{7}{8}=\frac{7}{8}$$

$$\frac{4}{8}+\frac{7}{8}=\frac{11}{8}$$

Similarly,

$$\frac{7}{8}-\frac{4}{8}=\frac{3}{8}$$

Multiplication of Fractions

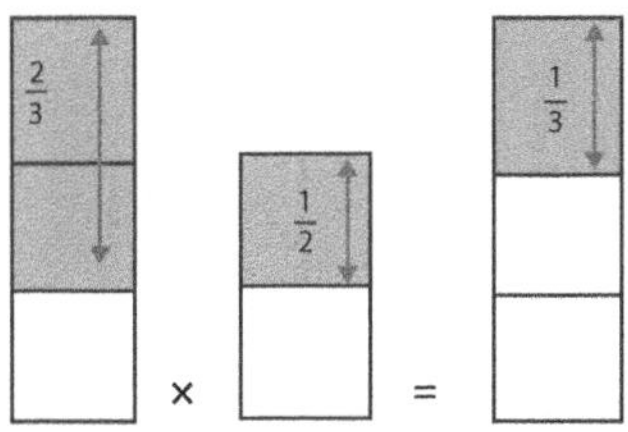

$$\frac{2}{3}\times\frac{1}{2}=\frac{2\times1}{3\times2}=\frac{2}{6}=\frac{1}{3}$$

***How to Multiply Fractions?**

- **Step 1:** Multiply the numerators.
- **Step 2:** Multiply the denominators.
- **Step 3:** Reduce the answer to the lowest terms.

For example:

$$\frac{1}{7}\times\frac{4}{6}=\frac{4}{42}=\frac{2}{21}$$

***How to Multiply Mixed Fractions?**

- **Step 1:** Convert the mixed numbers to improper fractions first.
- **Step 2:** Multiply the numerators.
- **Step 3:** Multiply the denominators.
- **Step 4:** Reduce the answer to the lowest terms.

For example:

$$2\frac{1}{3}\times 1\frac{1}{2}=\frac{7}{3}\times\frac{3}{2}=\frac{21}{6} \text{ which then reduces to } 3\frac{1}{2}$$

Division of Fractions

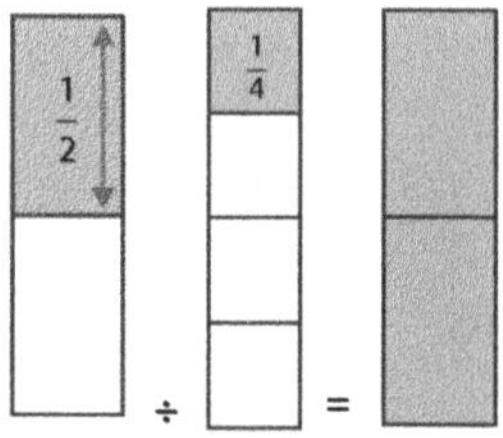

$$\frac{1}{2}\div\frac{1}{4}=\frac{1}{2}\times\frac{4}{1}=2$$

***How to Divide Fractions?**

- **Step 1:** Change the division sign to multiplication.
- **Step 2:** The fraction's inverse that follows the multiplication symbol (flip the fraction around so the old denominator is the new numerator, and the old numerator is the new denominator).
- **Step 3:** Multiply the numerators.
- **Step 4:** Multiply the denominators.
- **Step 5:** Change the answer to the lowest terms.

For example:

$$\frac{1}{8}\div\frac{2}{3}=\frac{1}{8}\times\frac{3}{2}=\frac{3}{16}$$

***How to Divide Mixed Fractions?**

- **Step 1:** Convert the mixed number to the improper fraction.
- **Step 2:** Reciprocate the dividend and multiply it with the divisor.
- **Step 3:** Multiply the numerators.
- **Step 4:** Multiply the denominators.
- **Step 5:** Change the answer to the lowest terms.

For example:

$$3\frac{3}{4} \div 2\frac{5}{6} = \frac{15}{4} \div \frac{17}{6} = \frac{15}{4} \times \frac{6}{17}$$

which, when solved, is $\frac{15 \times 6}{4 \times 17} = \frac{90}{68} = \frac{45}{34}$ which simplifies $1\frac{11}{34}$

***How to divide a whole number by a fraction, and a fraction by a whole number?**

- **Step 1:** Find the reciprocal of the fraction to be divided.
- **Step 2:** Multiply the number by the reciprocal of the fraction.
- **Step 3:** Simplify the resulting fraction if possible.
- **Step 4:** Check your answer: Multiply the result you got by the divisor and be sure it equals the original dividend.

*Note: *Division can be possible for only non-zero fractions.*
For example: to find a division of any whole number by a fraction.

$$3 \div \frac{5}{6} = 3 \times \frac{6}{5} = \frac{18}{5}$$

and to find a division of any fraction by a whole number

$$\frac{5}{6} \div 3 = \frac{5}{6} \times \frac{1}{3} = \frac{5}{18}$$

also, to find a division of any fraction by a fraction

$$\frac{3}{4} \div \frac{5}{6} = \frac{3}{4} \times \frac{6}{5} = \frac{18}{20} = \frac{9}{10}$$

EXAMPLE 4: (a) Shade $\frac{11}{18}$ (b) Shade $\frac{7}{18}$

SOLUTION:

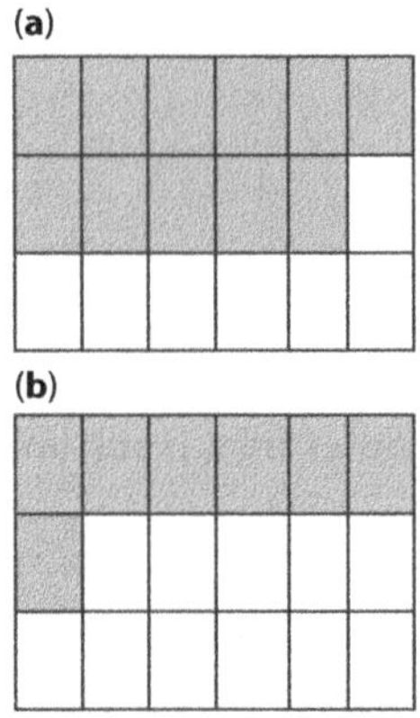

EXAMPLE 5: Write down three equivalent fractions of $\frac{11}{18}$
SOLUTION:

$$\frac{11}{18}=\frac{11}{18}\times\frac{2}{2}=\frac{22}{36}$$
$$\frac{11}{18}=\frac{11}{18}\times\frac{3}{3}=\frac{33}{54}$$
$$\frac{11}{18}=\frac{11}{18}\times\frac{4}{4}=\frac{44}{72}$$

As in all the three fractions $\frac{22}{36},\frac{33}{54},\frac{44}{72}$, the simplest form is $\frac{11}{18}$ hence we can say that they are equal or equivalent fractions.

EXAMPLE 6: Convert mixed fraction $5\frac{6}{7}$ to an improper fraction.
SOLUTION:

- **Step 1:** $5\frac{6}{7}=7\times5=35$
- **Step 2:** $6+35=41$
- **Step 3:** Leave the Denominator the same, i.e., 7.
- **Step 4:** The Improper fraction $=\frac{41}{7}$.

EXAMPLE 7: Convert mixed fractions $11\frac{1}{2}$ to improper fractions.
SOLUTION:

$$11\frac{1}{2}=\frac{(2\times 11)+1}{2}=\frac{(22+1)}{2}=\frac{23}{2}$$

EXAMPLE 8: Multiply the given Fractions $\frac{2}{3}$ and $\frac{5}{2}$
SOLUTION:

$$\frac{2}{3}\times\frac{5}{2}=\frac{2\times 5}{3\times 2}=\frac{10}{6}=\frac{5}{3}$$

EXAMPLE 9: Divide $3\frac{3}{4}$ by $2\frac{5}{6}$
SOLUTION:

$$3\frac{3}{4}\div 2\frac{5}{6}=\frac{15}{4}\div\frac{17}{6}=\frac{15}{4}\times\frac{6}{17}$$

which, when solved, is $\frac{15\times 6}{4\times 17}=\frac{90}{68}=\frac{45}{34}$ which simplifies $1\frac{11}{34}$

EXAMPLE 10: Divide $\frac{1}{8}$ by $\frac{2}{3}$
SOLUTION:

$$\frac{1}{8}\div\frac{2}{3}=\frac{1}{8}\times\frac{3}{2} \text{ which, when solved, is } \frac{3}{16}$$

Level 1: Exercise 1.1 Practice Questions

1. Color the given shapes to create the following improper fractions. Remember, each shape is one whole.

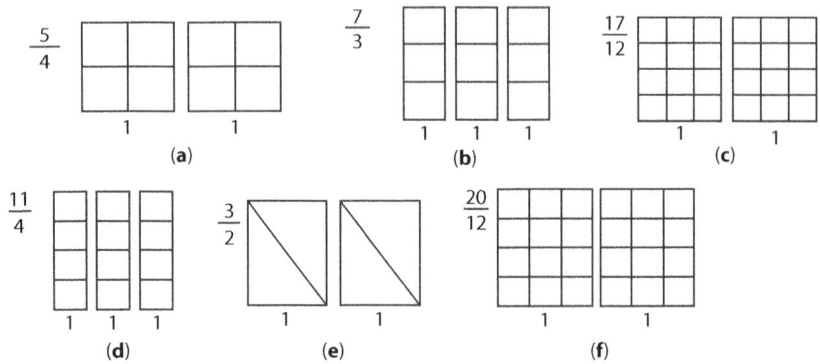

2. Write down three fractions equivalent to each of the following.

    a. $\frac{2}{3}$

    b. $\frac{4}{9}$

    c. $\frac{7}{10}$

3. Write these fractions in their simplest form.

    a. $\frac{6}{10}$

    b. $\frac{10}{40}$

    c. $\frac{24}{36}$

4. Convert to mixed fractions.

    a. $\frac{12}{5}$

    b. $\frac{24}{7}$

c. $\frac{41}{36}$

5. Convert to improper fractions.

   a. $1\frac{2}{5}$

   b. $2\frac{5}{6}$

   c. $3\frac{1}{4}$

6. Multiplying Fractions. Work out answers to the following.

   a. $\frac{1}{2} \times \frac{3}{5}$

   b. $\frac{2}{3} \times \frac{4}{7}$

   c. $\frac{3}{7} \times \frac{4}{5}$

7. Adding or Subtracting Fractions. Work out answers to the following.

   a. $\frac{3}{11} + \frac{7}{11}$

   b. $\frac{3}{4} + \frac{7}{8}$

   c. $\frac{4}{5} - \frac{1}{6}$

8. Convert these fractions to decimals.

   a. $\frac{5}{6}$

   b. $\frac{1}{3}$

c. $\frac{5}{13}$

Level 2: Exercise 1.2 Practice Questions

1. Compare the following set of fractions and put an appropriate sign < or >.

a. $\frac{5}{16} \underline{\qquad} \frac{2}{16}$

b. $\frac{3}{6} \underline{\qquad} \frac{1}{6}$

c. $\frac{1}{9} \underline{\qquad} \frac{6}{9}$

d. $\frac{8}{12} \underline{\qquad} \frac{5}{12}$

e. $\frac{3}{11} \underline{\qquad} \frac{5}{11}$

f. $\frac{4}{15} \underline{\qquad} \frac{2}{15}$

g. $\frac{2}{20} \underline{\qquad} \frac{4}{20}$

h. $\frac{1}{17} \underline{\qquad} \frac{13}{17}$

2. Convert the following fractions as improper fractions.

a. $1\frac{1}{4}$

b. $4\frac{1}{3}$

c. $1\frac{2}{3}$

d. $1\frac{3}{4}$

e. $1\frac{5}{4}$

f. $2\frac{1}{8}$

g. $1\frac{6}{37}$

h. $3\frac{4}{5}$

i. $7\frac{1}{4}$

j. $5\frac{2}{3}$

k. $3\frac{5}{6}$

l. $6\frac{1}{4}$

3. Convert the following fractions as mixed fractions.

a. $\frac{19}{17}$

b. $\frac{18}{7}$

c. $\frac{15}{11}$

d. $\frac{21}{11}$

e. $\frac{25}{17}$

f. $\frac{36}{15}$

g. $\frac{17}{8}$

h. $\frac{21}{4}$

i. $\frac{9}{4}$

j. $\frac{12}{5}$

k. $\frac{47}{17}$

l. $\frac{54}{7}$

4. Write in the lowest terms.

a. $\frac{14}{17}$

b. $\frac{21}{15}$

c. $\frac{18}{32}$

d. $\frac{12}{18}$

e. $\frac{48}{80}$

f. $\frac{9}{27}$

g. $\frac{13}{39}$

h. $\frac{10}{70}$

i. $7\frac{12}{21}$

j. $7\frac{11}{40}$

k. $5\frac{8}{14}$

l. $6\frac{10}{30}$

5. Find the missing number.

   a. $\frac{1}{17} = \frac{?}{34}$

   b. $\frac{5}{7} = \frac{?}{28}$

   c. $\frac{4}{7} = \frac{?}{35}$

   d. $\frac{5}{9} = \frac{?}{72}$

   e. $5\frac{2}{3} = \frac{?}{12}$

   f. $1\frac{4}{5} = \frac{?}{10}$

   g. $1\frac{1}{6} = \frac{?}{12}$

   h. $2\frac{3}{5} = \frac{?}{10}$

6. Convert the following fractions into decimals form.

   a. $\frac{2}{3}$

   b. $\frac{1}{8}$

   c. $\frac{8}{9}$

   d. $\frac{7}{13}$

   e. $\frac{6}{9}$

   f. $\frac{9}{16}$

7. Convert the following decimals to fractions form.
   a. 0.225
   b. 0.375
   c. 0.0175
   d. 0.95
   e. 0.5
   f. 0.45

8. Multiply the following fractions.
   a. $\frac{1}{9} \times \frac{1}{2} =$
   b. $\frac{7}{10} \times \frac{2}{5} =$
   c. $\frac{3}{8} \times \frac{2}{7} =$
   d. $\frac{4}{8} \times \frac{3}{7} =$
   e. $\frac{3}{4} \times \frac{2}{3} =$
   f. $\frac{7}{16} \times \frac{4}{3} =$
   g. $\frac{15}{64} \times \frac{1}{12} =$
   h. $\frac{2}{3} \times \frac{5}{9} =$
   i. $\frac{3}{4} \times 10 =$
   j. $1\frac{1}{2} \times \frac{5}{6} =$
   k. $\frac{3}{16} \times \frac{5}{12} =$
   l. $14 \times \frac{3}{8} =$

m. $\frac{1}{2} \times 1\frac{1}{3} =$

n. $3\frac{1}{16} \times \frac{1}{5}$

o. $18 \times 1\frac{1}{2} =$

p. $16 \times \frac{2}{8} =$

q. $6\frac{3}{8} \times 1\frac{3}{5} =$

r. $2\frac{1}{2} \times 4\frac{1}{3} =$

s. $5\frac{7}{9} \times 6\frac{1}{7} =$

t. $4\frac{1}{2} \times 6\frac{1}{5} =$

9. Divide as shown.

a. $\frac{1}{4} \div \frac{2}{3} =$

b. $\frac{3}{5} \div \frac{2}{7} =$

c. $\frac{3}{7} \div \frac{2}{3} =$

d. $\frac{2}{9} \div \frac{1}{3} =$

e. $4 \div \frac{2}{3}$

f. $8 \div \frac{2}{7} =$

g. $9 \div \frac{5}{3} =$

h. $\frac{3}{4} \div \frac{4}{3} =$

i. $\frac{4}{11} \div \frac{1}{11} =$

j. $\frac{3}{4} \div \frac{8}{11} =$

k. $\frac{1}{4} \div 7 =$

l. $4 \div \frac{7}{8} =$

m. $45 \div \frac{1}{3} =$

n. $8 \div \frac{3}{4} =$

o. $1\frac{3}{4} \div 1\frac{2}{3} =$

p. $2\frac{7}{9} \div 3\frac{1}{7} =$

q. $2\frac{1}{2} \div 4\frac{1}{3} =$

r. $5\frac{1}{4} \div 2\frac{2}{3} =$

s. $\frac{3}{4} \div 1\frac{5}{9} =$

t. $3\frac{3}{4} \div 4\frac{2}{3} =$

10. Add or subtract as shown.

a. $\frac{1}{9} + \frac{1}{2} =$

b. $\frac{7}{10} + \frac{2}{5} =$

c. $\frac{3}{8} + \frac{2}{7} =$

d. $\frac{4}{8}+\frac{3}{7}=$

e. $\frac{3}{4}-\frac{2}{3}=$

f. $\frac{7}{16}-\frac{4}{3}=$

g. $\frac{15}{64}-\frac{1}{12}=$

h. $\frac{2}{3}-\frac{5}{9}=$

i. $\frac{3}{4}+10=$

j. $1\frac{1}{2}+\frac{5}{6}=$

k. $\frac{3}{16}+\frac{5}{12}=$

l. $14+\frac{3}{8}=$

m. $\frac{1}{2}-1\frac{1}{3}=$

n. $3\frac{1}{16}-\frac{1}{5}=$

o. $18-1\frac{1}{2}=$

p. $16-\frac{2}{8}=$

q. $6\frac{3}{8}+1\frac{3}{5}=$

r. $2\frac{1}{2}+4\frac{1}{3}=$

s. $5\frac{7}{9}+6\frac{1}{7}=$

t. $4\frac{1}{2}+6\frac{1}{5}=$

Level 3: Exercise 1.3 Practice Questions

1. Throughout his workout, Sameer burns 450 calories every half-hour. How many calories will Sameer burn if he exercises for 1¾ hours at this rate?
2. Adam spent $7\frac{1}{2}$ hours studying on Friday. Write this period as a fraction of the day.
3. Sarah purchased a packet of cupcakes at the store. On her way home, she ate ¼ cupcakes from the packet. Later that night, she served ½ of what was left at dinner with friends. If six cupcakes are left in the pocket, how many did the packet contain?
4. Sameer's Electrical shop has a play station on sale for $\frac{1}{3}$ off the regular price of $429. What is the sale price of the play station?
5. For a family party, Aisha made $\frac{2}{5}$ dishes from all the dishes made on that day. If 40 dishes were made for the party, how many did Aisha make?
6. The price of mobile has been reduced by $\frac{2}{5}$. If the price of a mobile was originally $1027, how much has the price reduced?

Mental Maths trick:
Addition of two fraction numbers
Step1: - Multiply crosswise and add them to get the numerator of the answer.
Step2: - Multiply the bottom (denominators) of two fractions to get the bottom of the answer.
Example:

$$\frac{5}{3}+\frac{7}{4}=\frac{20+21}{12}=\frac{41}{12}$$

Here 5×4=20 and 7×3=21. Then 20+21=41. The bottom of the fraction is just 3×4=12. So, the answer is 41/12.
Subtraction of two fraction numbers (same as that of addition)
Step1: - Multiply crosswise and subtract them to get the numerator of the answer.
Step2: - Multiply the bottom (denominators) of two fractions to get the bottom of the answer.

Example:

$$\frac{5}{3}-\frac{7}{4}=\frac{20-21}{12}=\frac{-1}{12}$$

Here 5×4=20 and 7×3=21. Then 20-21= -1. The bottom of the fraction is just 3×4=12. Therefore, the answer is -1/12.

1.3 Decimals and Percentages

Decimal numbers have a decimal point separating the complete number from the fractional portion. The dot (.) that occurs between the components of a whole number and a fraction is the decimal point.

An example of a decimal number is '21.2' In this case, 21 is a whole number, and 2 is the fractional component. The decimal point is "."

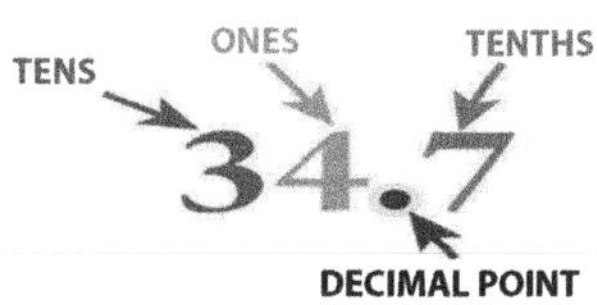

21.2 has 2 Tens, 1 One, and 2 Tenths

For example: 3.2376 (Finite), 3.137654….(Infinite)

A percentage describes 100 parts, which means how many parts are out of one hundred parts of a particular thing. If 25% of the people in a theatre wear red shirts, then for every 100 people in the theatre, 25 of them wear red shirts.

How to convert decimals and fractions?

We need to use the place value of the last digit to convert a decimal to a fraction.

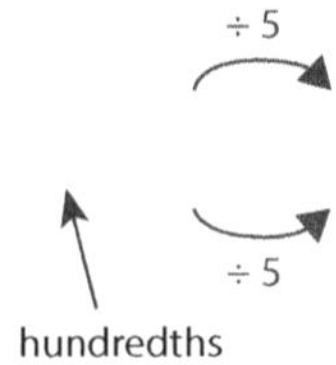

For example:

$$0.85 = \frac{85}{100} = \frac{17}{20}$$

To convert a fraction to a decimal: Divide the bottom number from the top number.

For example:

$$\frac{4}{5} = 4 \div 5 = 0.8$$

How to convert percentages and fractions or decimals?

Divide by 100 to write a % as a fraction or decimal:

For example:

$$64\% = 64 \div 100 = 0.64$$

or

$$64\% = \frac{64}{100} = \frac{16}{25}$$

Multiply by 100 to write a decimal or fraction as a %:

For example:

$$0.125 = 0.125 \times 100 = 12.5\%$$

$$\frac{2}{5} = \frac{2}{5} \times 100 \left(\text{i.e.} \frac{2}{5} \text{ of } 100\% \right) = 40\% \quad \text{Or, } \frac{2}{5} = 2 \div 5 \times 100 = 40\%$$

The formula for Percent Increase/Decrease

$$\textbf{percent increase} = \frac{\textit{amount of increase}}{\textit{original amount}}$$

OR

$$\textbf{percent decrease} = \frac{\textit{amount of decrease}}{\textit{original amount}}$$

Level 1: Exercise 1.4 Practice Questions

1. Convert these decimals as fractions:
   a. 0.3 =
   b. 0.5 =
   c. 0.6 =
   d. 0.02 =
   e. 0.05 =
   f. 0.25 =
   g. 0.36 =
   h. 0.125 =

2. Convert these fractions as decimals:
   a. $\frac{7}{10} =$
   b. $\frac{1}{5} =$
   c. $\frac{2}{5} =$
   d. $\frac{6}{11} =$
   e. $\frac{3}{4} =$
   f. $\frac{5}{4} =$
   g. $\frac{3}{7} =$

h. $\frac{3}{20} =$

i. $\frac{3}{25} =$

3. Convert these percentages as decimals:
   a. 3% =
   b. 30% =
   c. 25% =
   d. 8% =
   e. 12% =
   f. 67% =

4. Convert these percentages as fractions:
   a. 20% =
   b. 75% =
   c. 5% =
   d. 30% =
   e. 40% =
   f. 15% =
   g. 24% =
   h. 35% =

5. Convert these decimals as percentages:
   a. 0.25 =
   b. 0.5 =
   c. 0.7 =
   d. 0.07 =
   e. 0.45 =
   f. 0.09 =
   g. 0.4 =
   h. 0.375 =

6. Convert these fractions as percentages:

   a. $\frac{3}{10} =$

   b. $\frac{3}{9} =$

c. $\frac{3}{17} =$

d. $\frac{3}{4} =$

e. $\frac{4}{7} =$

f. $\frac{13}{40} =$

g. $\frac{3}{12} =$

7. Calculate the percentage of the given Quantities:

    a. 23% of 75 is what number?
    b. What number is 72% of 45?
    c. 40 is what percent of 80?
    d. What percent of 88 is 8?
    e. 12 is 52% of what number?
    f. 16% of what number is 48?
    g. What number is 73% of 150?
    h. 62 is what percent of 58?
    i. 37 is 140% of what number?
    j. What percent of 678 is 65?
    k. 98% of what number is 17?
    l. 0.4% of 49 is what number?

Level 2: Exercise 1.5 Practice Questions

1. Because of the popularity of online shopping, about 50 of the 300 traditional retail shops across city may close in the next few years. Find the percentage of retail shops that may close.
2. In a class, half of the students have iPhones, 30% of the class have Samsung phones, and the remaining do not have any smartphones. Express the fraction of the students who don't have a smartphone.

3. Aisha went to the market to buy groceries with her mom. They purchased 2 kg of potatoes and 18 kilograms of rice. What is the percentage of the weight of potato and the weight of rice?
4. A furniture designer is making a piece of furniture. He needs six pieces of wood 200 mm long. What fraction of this piece will be leftover if he cuts them from 3,000 mm long?
5. Sarah cut her birthday cake into 8 equal slices and ate 2 of them straight away. What fraction was left?
6. Sarah bought a dress that was on sale for $24.50 after a discount of 25%. Calculate the original price of the dress.
7. The weight of a bicycle is 20% of a car's weight. If a bicycle weighs 15 kg, how much does a Car weigh?
8. I love chocolate. My mum buys a big bar and says I can have 1/8, 10%, or 0.12 of the bar. Which of these will give me the most chocolate?
9. A bike is on sale for $1700, which is 80% of the original price. Find the actual price.
10. During the summer holidays, 146 students out of the 680 attended classes on Sunday. Calculate the percentage of absent students.

Level 3: Exercise 1.6 Practice Questions

1. In 2023, Adam deposited $ 2,350 into his bank account. One year later, this amount had increased by 25%. How much did he have in his account in 2024?
2. Sameer's new record for the 100m running is 11.8 seconds, which is 4.8% faster than his previous record. What was his previous record? (Write your answer up to 2dp)
3. Sameer buys a car. After two years of owning it, it is now worth $ 10,120, constituting a 56% decrease from its original value. What was its initial value?
4. Mr. Adam has a monthly salary of $ 850. If he spends $ 525 monthly, find his spending and reserves in percentage.
5. In a university, there are 1200 boys, and some are girls. If the number of boys is 60 % of the total number of students at university, calculate the number of girls.
6. Sameer bought a secondhand car and spent 25 % of the buying cost on repair. If his total cost, including repairs, was $ 1500, at what price did he buy the car?

7. The total holding capacity of a water tank is 250 liters, and it is 70 % full. Some water is taken from the tank, and it is found that the tank is now 45 % full. Find the quantity of water taken.
8. In a mixture of two-color paints (Red and Yellow), 35% is yellow Paint. If the total quantity of the mix is 20 kg, find the amount of red paint, by weight, in the mixture.
9. The price of a smartphone has increased from \$870 to \$ 920; find the percentage increase.
10. After an increase of 25 %, the number becomes 840. Find the original number.
11. Mr. Adam saves 38 % of his income; what percent of his income does he spend?
12. Sarah's score on the test was 84. Her score on the second test was 92. What was the percentage of the increase in her score?
13. A football is on sale for BHD32. The original price was BHD58. What is the percent decrease in the price?
14. A project team has 120 members. Of the 120 members, 25% bought their air tickets at the rate of \$250, and the remaining bought at the rate of \$230, how much money does the whole team spend altogether on air tickets.

1.4 Check Your Knowledge (MCQ)

1. What the product of $\frac{1}{2}$ and $\frac{2}{5}$?

   a. $\frac{2}{5}$

   b. $\frac{4}{5}$

   c. $\frac{3}{5}$

   d. $\frac{1}{5}$

2. Which of the following is equal to $\frac{3}{8} \times 2\frac{2}{7}$?
   a. $\frac{21}{128}$
   b. $\frac{3}{14}$
   c. $\frac{6}{7}$
   d. $\frac{3}{2}$

3. Which is a decimal representation of 10.8% ?
   a. 0.108
   b. 1.08
   c. 10.8
   d. 108

4. Which is percentage representation of 1.25 ?
   a. 0.125%
   b. 1.25%
   c. 12.5%
   d. 125%

5. 25 students participate in their college's sports programs. 11 of the students are male. What decimal number best captures the proportion of male students?
   a. 0.44
   b. 0.52
   c. 0.48
   d. 0.56

6. It rained 40% of the days during April. How many days has it rained?
   a. 20
   b. 15
   c. 12
   d. 10

7. 500 tourists in the area said that 31% of them were fearful about participating in water activities. How many individuals reported being fearful about participating in water sports?
   a. 31
   b. 155
   c. 500
   d. 200

8. Which statement is true?
   a. 0.37>59%
   b. 59%<0.53
   c. 75%>0.8
   d. 0.095>5.9%

9. Convert the decimal number 0.05 as a percentage.
   a. 0.05%
   b. 0.5%
   c. 5%
   d. 50%

10. During her first training day, Aisha completes 2 hours of running. Which digit represents the duration of her run?
    a. 0.08 hr
    b. 0.22hr
    c. 0.8hr
    d. 0.02hr

1.5 Activity/Experiment

Activity 1: Online Shopping!

There is an online sale in Sharaf DG Bahrain with some percentage of the discount. You are then allowed BHD 500 (Bahraini Dinar) to purchase as many items on sale as possible.

Answer these questions.
- How much did you pay for each item?
- How much money did you save?
- How much money had you leftover?

Activity2

Often you can see percentages in shops when it is sale time. Work out the sale price of these items:

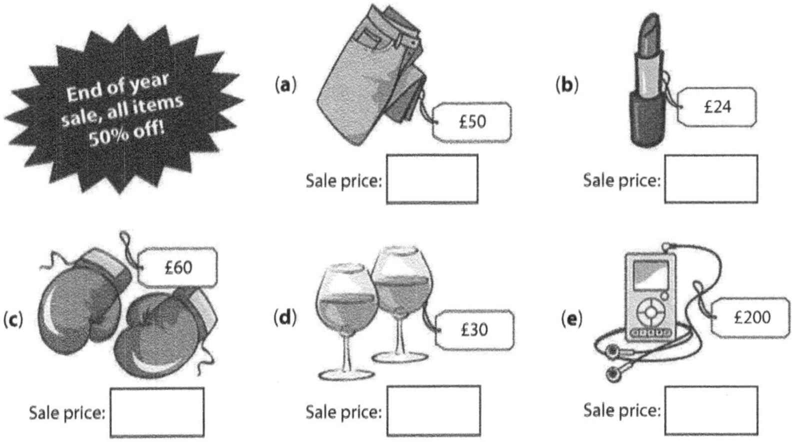

2
Ratio and Proportion

Objectives: This chapter

- *Demonstrate how to express ratios as fractions and rates as fractions.*
- *Determine proportions.*
- *Solve problems by writing proportions.*
- *Apply ratios and proportions to solve real-life problems.*

2.1 Introduction

Every day in our lives, we need to compare quantities. We use the ratio to compare every field, such as homes, businesses, sports, science, engineering, etc.

There are a few examples of the ratio.

- The football team won 5 games for every 3 games lost.
- The employee gets a three-monthly bonus every two years.

In simple words, we can say that ratios are another way of demonstrating fractions.

Ritu Shrivastava, Ramakant Bhardwaj and Satyendra Narayan. Mathematics for Engineers, (37–58)

2.2 Ratios

2.2.1 Definition

The relationship between two quantities is called a ratio and is generally expressed as the quotient of one divided by the other.

Note 1: *To find any ratio, do we need to ensure that the quantities are in the same unit? If not, we need to convert it by using an appropriate rule.*
Note 2: *We can write ratio in three different ways. like the ratio of 1 to 2 can be written as*

> *1 is to 2 or 1: 2 or* $\frac{1}{2}$
> *Remember, no matter which notation we use for this ratio; it is read as "1 is to 2."*

2.2.2 Simplifying Ratio

We already learned about the simplest form of any fraction in the previous chapter.

> *If a fraction's numerator and denominator share only one common factor, it is said to be in its simplest form.*

The same will apply to the ratio.
For example: Write the ratio 50 to 100 in the simplest form.

SOLUTION: The ratio 50 to 100 can be expressed as a fraction $\frac{50}{100}$

$$\frac{50}{100} = \frac{50/50}{100/50} = \frac{1}{2}$$

So, the ratio of 50 to 100 is the same as that of 1 to 2.

2.2.3 Equivalent Ratio

The ratio of 3:2 can also be written as 6:4.

Two ratios are equivalent if the fraction corresponding to them is equivalent.

For example: $\frac{1}{3}$ to $\frac{2}{6}$ are equivalent ratios since they represent the same fraction.

2.2.4 Methods for Finding a Ratio

Unitary method: In this method, first, we find the value of one unit or the value of one part and then calculate the value of the required number of units or parts.

Problem type 1: Where ratio and total amount are given.
EXAMPLE: Divide BHD 16 into 3: 5.

- **Step 1:** Let the two ratios be $3x$ and $5x$.
- **Step 2:** Given that the total ratio is equal to BHD 16.
- **Step 3:** Therefore, $2x + 7x = 18 = 9x = 18$

$$x = \frac{18}{9} = 2$$

- **Step 4:** So that the value of one-part equals 2 BHD. Multiply by the ratio numbers

$$3 \times 2 = 6$$
$$5 \times 2 = 10$$

- **Step 5:** Check your answers. Values from Step 4 should be added to the correct amount. 6+10 = 16

> ***Scholar's style:*** *Total given*
>
> *Total ratio = 3+5 = 8*
> *Value of one part = = 2 BD*
> *Value of 3 parts = 3 × 2 = 6BD*
> *Value of 5 parts = 5 × 2 = 10BD*
> *Total 6BD + 10 BD = 16BD*

Problem type 2: Where total is not given, but the ratio and the amount of any one ratio are provided.
EXAMPLE: Divide the money into two parts in the ratio of 3:7. If the large part is BHD 735, find the total amount.

- **Step 1:** Let the two ratios be $3x$ and $7x$.
- **Step 2:** Given that $7x$ (larger part) = 735
- **Step 3:** Therefore, $x = \frac{735}{7} = 105$
- **Step 4:** So that the value of one-part equals 105.

Multiply this value with the remaining ratio $3 \times 105 = 315$

- **Step 5:** Add up each amount to get Total = 735 + 315 = 1050 BHD

Therefore, the total amount is =1050 BHD

> ***Scholar's style:*** *Total = unknown*
>
> *Given ratio 3: 7*
> *Given the value of 7 parts 7x= 735*
> *So the value of one part x = 105*
> *Multiply this value by the remaining ratio*
> *3 × 105 = 315 (value of 3 parts)*
> *Add up both amount = 315 + 735 = 1050 BD*

2.2.5 Methods to Calculate Increasing and Decreasing in Each Ratio

For any given ratio, it is easy to calculate the increase and the decrease by just multiplying or dividing the same number of both ratios, have a look at the given example.

In a recipe for tea for eight people, we need 2 cups of milk, 4 spoons of sugar, 4 spoons of tea powder, ½ spoon of cardamom powder and ½ spoon of ginger powder. How much of each component is required to make tea for 24 people using the ratio of ingredients provided?

SOLUTION: Find what each amount needs to be multiplied (or divided) by. In this case, tea for 8 people must be made for 24 people. Since $8 \times 3 = 24$

So, everything needs to be multiplied by 3.
Step 2:

Ingredients	**8 people**	**24 people**
Milk	2	6
Sugar	4	12
Tea powder	4	12
Cardamom	½	1 ½
Ginger powder	½	1 ½

EXAMPLE 1: Express the ratio 90: 180 in the simplest form.
SOLUTION: To express any ratio in its simplest form, we need to divide both ratios by their HCF (highest common factor)
Let's find the HCF of both ratios.
Factors of 90 are 1,2,3,5,6,9,10,15,18,30,45,90.
Factors of 180 are 1,2,3,4,5,6,9,10,12,15,18,30,36,45,60,90, 180.
We can observe that 90 and 180 have a 90 HCF (Highest Common Factor).
Therefore, $\frac{90}{180} = \frac{90/90}{180/90} = \frac{1}{2}$
Hence the simplest form of 90 and 180 is 1:2.

EXAMPLE 2: Sameer and Sarah got 6 and 8 marks on a math test. What is the ratio of their marks in the simplest form?
SOLUTION: Given that, the ratio of marks = 6:8
Let's find the HCF of both ratios.
Factors of 6 are 1,2,3,6
Factors of 8 are 1,2,4,8
The Highest Common Factor (HCF) of 6 and 8 is 2.
Therefore, $\frac{6}{8} = \frac{6/2}{8/2} = \frac{3}{4}$
Hence the simplest form of Sameer and Sarah's marks in the math's test is 3:4.

EXAMPLE 3: Divide 660 cm into 1: 2: 3.
SOLUTION: Add up the ratio number → 1+2+3 = 6

Divide the total by this number → $\frac{660}{6} = 110$

Multiply by the ratio number.

$$1 \times 110 = 110 \text{ cm}$$
$$2 \times 110 = 220 \text{ cm}$$
$$3 \times 110 = 330 \text{ cm}$$

Check that your answer adds to the correct number.

$$110 + 220 + 330 = 660 \text{ cm}$$

EXAMPLE 4: In a triangle, all three angles are in a ratio of 1:2:3. Suppose the largest angle is 90°. Find the other two angles.
SOLUTION: Divide the given angle by the corresponding ratio

$$\frac{90}{3} = 30^\circ$$

Multiply the value with each ratio $1 \times 30^\circ = 30^\circ$

$$2 \times 30^\circ = 60^\circ$$
$$3 \times 30^\circ = 90^\circ$$

Add up all angles to check your answer 30° + 60° + 90° = 180°

EXAMPLE 5: Three numbers are in a ratio of 2:3:5. If the total of these three numbers is 200, find the numbers.
SOLUTION: Since the given ratio is 2:3:5. So the three numbers are $2x$, $3x$, and $5x$,
Thus, we must divide 200 into three parts, $2x$, $3x$, and $5x$.
Therefore, $2x + 3x + 5x = 200$
$10x = 200$

$$x = \frac{200}{10} = 20$$

$$\text{So the first number} = 2x = 2 \times 20 = 40$$

$$\text{Second number} = 3x = 3 \times 20 = 60$$

$$\text{The third number} = 5x = 5 \times 20 = 100$$

To check your answer, add all numbers 40 + 60 + 100 =200.

Level 1: Exercise 2.1 Practice Questions

1. In a mall, the ratio of female to male students is 15:25; express the ratio in its simplest form.
2. Aisha did 25 hours of work, and Sana did 75 hours of work. Find the ratio of Aisha's work to Sana's work.
3. Write the following ratios in their simplest form.
   a. 12: 48
   b. 9: 36
   c. 3: 6: 12
   d. 55: 77
   e. 500 L: 4000 L
   f. 70 kg: 350 kg
   g. 3: 21: 90
   h. 6 km: 1250 km
   i. 16 ml: 3 L
   j. 1600 km: 2 km
   k. BHD 7: 7000 fills [1 BHD = 1000fills]
   l. (l) 1000 g: $\frac{1}{2}$ kg
4. In a mathematics assessment, Aisha and Sana got their grades in the ratio of 2:3, and the sum of their marks was 225. Find how many marks each of them got.
5. Kevin wants to divide BHD108 in the ratio of 2:7 for his friends Sameer and Sameera. How much money will each get?
6. Total weight distributed into the three baggage in the ratio 3:4:8. Find the weight of each bag.
7. In a quiz competition, the total prize money of BHD 828 is to be distributed among the first, second, and third place winners in the ratio 5: 3: 1. Calculate how much each one will get.
8. Sarah and Aisha eat pizza in a ratio of 2:3. If there are ten pieces in the pizza, how much should they each get?

9. The syrup recipe requires water and sugar in the ratio of 5:2. Suppose the total amount of sugar is 14 L(liter). How much water is needed?
10. A paint mix has red and green paint in the ratio of 3:7. Suppose 2500ml of paint is to be made up. How much paint should there be?

Level 2: Exercise 2.2 Practice Questions

1. A herd has 15 White and some Block Horses. The ratio of white horses to black Horses is 3:5. Calculate how many Black Horses are in the pack.
2. Sana and Aisha shared a reward of BHD x in a ratio of 2:5. If Sana received BHD 40, find the total amount as a reward.
3. Sixty-six girls participated in a swimming competition. The ratio between girls to boys was 6:5. Find out how many total members participated in the competition.
4. A fruit box has apple, Banana, and Mango fruits in a ratio of 3:1:3. If there are 21 Mangos in the box, how many fruits are there?
5. Two numbers have a ratio of 3:5. The larger number is 50, less than twice the smaller one. Find the numbers.
6. A piece of wire is cut into three parts in a ratio of 3:5:7. If the biggest part is 42cm, find the total length of the wire.
7. The monthly expenses and the monthly savings of a family are in a ratio of 6:5. If the savings are $500, find the expenses and the total income.
8. If the ratio of the cost of the notebook to the book is 1:10, if the notebook is $2.5, find the cost of the book.
9. Sana, Aisha, and Alya shared candies in a ratio of 4:5:7. Alya got 15 more candies than Sana. Work out the total number of candies they shared.

Level 3: Exercise 2.3 Mixed Practice Questions

1. If the ratio of length to width of a rectangular field is 5:2, and its perimeter is equal to 336 meters. Calculate the total area of the rectangular field.
2. Suppose each side of Yellow Square is twice as long as each side of Red Square. Determine the difference between Yellow Square's and Red Square's perimeters.

3. At the start of the semester, the library of a university had M1 and M2 mathematics books in the ratio of 5:3. By the end of the semester, 30% of each type of book were issued by the students, and 2240 books of both types remain in stock of library. Find the number of books of every kind in the library at the start of the semester.
4. What must be added to each term of the ratio of 3:2 so that it may become equal to 5:4?
5. Aisha and Sarah are in the ratio of 2:3. Twelve years ago; their ages were in the ratio of 8:13. What is their age now?
6. A bag contains $ 280 in the denominations of $ 20, $10, and $1 in the ratio of 3:4:7. Find the number of each denomination.
7. Aisha had 30 candies in her bag, 18 red and 12 yellow. Sameer has 15 candies in his bag in red and yellow colors. If the ratio of red candies to yellow candies is the same for Aisha and Sameer, then Aisha had how many more yellow candies than Sameer?
8. Mathematics class starts at the beginning of semester 1, with 12 boys and 8 girls in the class. After two weeks, 6 new boys and 4 girls joined the same class. Will the ratio of boys to the total number of students in the class remain the same? Explain.

2.3 Rate

Comparing different units or different kinds of measurement of unlike quantities is called a "Rate." A "Rate" is a ratio of, unlike quantities.

EXAMPLE 6: Suppose that you pay BHD 100 for every 8 hours of work.

$$\frac{BD\,100}{8\,hours} = \frac{BD\,25}{2\,hours}$$

So, you paid BHD 12.5 for every hour that you worked. Note that we must express the answer in the form of units.

EXAMPLE 7: Simplify the rate from 300 km to 18 L
SOLUTION: 300 km to 30 L

$$= \frac{300\ km}{18\ L} = \frac{50\ km}{3\ L}$$

2.3.1 Unit Rates

A unit rate is when the number in the denominator is 1.

EXAMPLE 8: Write a unit rate of 275 km in 5 hours.
SOLUTION: 275 km in 5 hours

$$\frac{275\ km}{5\ hours} = \frac{55\ km}{1\ hours}$$

2.4 Proportions

We can say two equal ratios are known as proportion.

Consider the two ratios, 3:12 and 2:8

$$3:12 = \frac{3}{12} = \frac{1}{4}$$
$$2:8 = \frac{2}{8} = \frac{1}{4}$$

Therefore 3:12 = 2:8 is a proportion.

a: b = c: d is a way to show proportion, which can be read as "a" is to "b" as "c" is to "d."
where a, b, c, and d are any numbers, the ratio of the first two numbers (a and b) equals the ratio of the last two numbers (c and d).

The first and the fourth terms of a proportion (a and d) are called ***extreme*** or ***extremes.***

The second and third terms (b and c) are the ***middle terms or means.***

Note: *If any four numbers are in the proposition, then the Product of the extreme terms equals the Product of the middle terms.*

EXAMPLE 9: Check these terms and whether 15,20,45,60 is in proportion or not.

SOLUTION: Given that the four terms are 15, 20, 45, and 60.

$$\text{Since } 15:20 = \frac{15}{20} = \frac{3}{4}$$

$$\text{And } 45:60 = \frac{45}{60} = \frac{3}{4}$$

Hence 15:20 = 45:60

Therefore, these four terms are in proportion.

Also, we can check by using the rule (Product of the extreme terms as equals to the Product of the middle terms).

$$\text{Product of extremes} = 15 \times 60 = 900$$

$$\text{Product of means} = 20 \times 45 = 900$$

Therefore, the product of extremes = product of means

Hence, the given four terms are in proportion.

2.4.1 Direct Proportion

When we observe a pattern of decrease -decrease or increase- increase while comparing any two quantities, we consider this situation as a direct proportion. It means when one quantity is increasing other is also increasing, and when one quantity is decreasing other is also decreasing.

As a definition, we can say that.

> "When two quantities *increase or decrease* in the same ratio, they are said to be in direct proportion."

Suppose you increase the number of clothes in your travel bag. What will happen to its weight?

It will also increase. This is called a direct proportion. In our day-to-day lives, everywhere we use proportions. Below three laws involves direct proportion, which is used in engineering.

(1) **Hooke's law:** states that, within the elastic limit of a material, the strain ε produced is directly proportional to the stress σ producing it, that is $\varepsilon \propto \sigma$. where '$\propto$' means "is proportional to"

(2) **Charles's law:** explains that in each amount of gas at constant pressure, the volume V is directly proportional to its thermodynamic temperature T. That is, $V \propto T$.

(3) **Ohm's law:** states that the current I, flowing through a fixed resistance, is directly proportional to the applied voltage V. That is, $I \propto V$.

EXAMPLE 10: Sameer can read 3 books in 15 days. How many similar books can he read in 30 days?

SOLUTION: As we know, two quantities are in direct proportion when they increase or decrease in the same ratio.

Let's compare the two given quantities in the above example.

Here we can identify that if Sameer reads for more days, he can read more books.

So, this is a direct proportion.

Let's solve this problem.

Let in 20 days, Sameer can read x books.

Since $3:15::x:30$

Product of extremes $3 \times 30 = 90$

The product of means $15 \times x = 12x$

Therefore, $15x = 90$

$$x = \frac{90}{15}$$

$$x = 6$$

Therefore, he can read 6 books in 30 days.

Also, we can use the ***Unitary method*** to solve this problem,

In 15 days, Sameer can read 3 books.

Therefore, in 1 day, he can read $\frac{3}{15}$ books.

Hence, in 30 days, he can read $\frac{3}{15} \times 30$ books.

He can read 6 books in 30 days.

Scholar's style:

Step 1: Write both quantities separately.

	Book	Day
Given	*3*	*15*
Let	*x*	*30*

Step 2: Cross-multiply the above quantities.

$\Rightarrow \quad 3 \times 30 = 15 \times x$

$\Rightarrow \quad 90 = 15x$

$\Rightarrow \quad x = 6$

Therefore, Jasim will finish approximately 6 books in 30 days

2.4.2 Inverse Proportion

Two quantities are in Inverse proportion; when one increases, the other will decrease, and vice versa.

Suppose you increase the speed of your car; what result will it be? You will reach your destination fast.

So, here we can absorb that while speed is increasing, time will decrease. This is called inverse proportion.

EXAMPLE 11: How many workers will be needed to do the same task in 30 hours if 20 workers can construct a wall in 48 hours?

SOLUTION: As we know, two quantities are in Inverse proportion when they increase or decrease in a different ratio.

Let's compare two quantities in the given example here. We can see that if 20 workers can build a wall in 48 hours, more workers will be required to finish work in 30 hours.

So, this is a problem of inverse proportion.

We can use the ***Unitary method*** to solve this problem.

48 hours are required to build a wall with 20 workers.

1 hour is required to build a wall by (20× 48) workers. [less hours, more workers

30 hours required to build a wall by (20 × 8)/20 workers [more hours, less workers] = 32 workers

Hence, 32 workers will be required to do the same work in 30 hours.

Scholar's style:

Step 1: Write both quantities separately

	Workers		Hours
Given	20	$\longrightarrow$	48
Let	x	$\longrightarrow$	30

Step2: Multiply both sides directly

$20 \times 48 = 30 \times x$

$x = 32$ *workers*

EXAMPLE 12: Are 32, 42, 52, and 62 in proportion?
SOLUTION: The four terms are 32, 42, 52 and 62
The product of extremes = 32 × 62 = 1984
and the product of means = 42 × 52 = 2184
Hence, the given four terms are not in proportion.

EXAMPLE 13: If 2, x, 12, and 30 are in proportion, find the value of x.
SOLUTION: Given that the four terms 2, x, 12, and 30 are in proportion.
Therefore, product of extremes $= 2 \times 30 = 60$

$$\text{Product of means} = x \times 12 = 12x$$
$$60 = 12x$$
$$x = \frac{60}{12}$$
$$x = 5$$

EXAMPLE 14: If the cost of 3 books is \$18, find the cost of 8 books of this kind.
SOLUTION: Since the given problem is a direct proportion.
Let the cost of 8 such books =\$ x
Since 3: 18:: 8: x
Product of extremes 3 × x = $3x$
Product of means 18 × 8 = 144
Therefore, $3x$= 144

$$x = \frac{144}{3}$$
$$x = 48$$

Therefore, the cost of 8 such books will be \$48.

EXAMPLE 15: A plane flies 235 km in half an hour. At this rate, how far can it fly in 1.5 hours?

SOLUTION: Since the given problem is a direct proportion.

Let the plane can fly up to x km in 1.5 hours.

Since 235: 30:: x: 90

(1 hr. = 60 min = ½ hour = 30 minutes)

(Similarly, = 1.5 hours = 60 + 30 = 90 minutes)

Therefore, the Product of extremes 235 × 90 = 21150

Product of means 30 × x = 30x

Therefore, 30x = 21150

$$x = \frac{21150}{30} = 705$$

Hence, the plane can fly up to 705 km in 1.5 hours.

EXAMPLE 16: 8 programmers working 7 hours a day can complete a program in 16 days. How many days will 4 programmers take to do the same job?

SOLUTION: First, we need to determine which kind of proportion it is.

8 programmers working 7 hours a day can finish the job in 16 days. Will less programmers take more or less time to do the job? less programmers means they'll take more time to complete the job. So, we're dealing with the inverse proportion. By using the ***Unitary method***

8 programmers working 7 hours a day can finish the job in 16 days.

8 programmers working 1 hour a day can finish it in (16 × 7) days [less hours per day, more days]

1 programmer working 1 hour a day can finish it in (16 × 7 × 8) days [less typists, more days]

4 programmers working 1 hour a day can finish it (16 × 7 × 8)/4 days [more typists, less days]

4 programmers working 7 hours a day can complete it $(16\times7\times8)/(4\times7)$ days [more typists, less days]= 32 days.

Hence, 4 programmers working 7 hours a day can finish the job in 32 days.

EXAMPLE 17: A house had enough groceries for 30 family members for 9 days. After 2 days, 5 family members left the house for a trip. If the groceries will be consumed at the same rate, calculate the number of days it would take to finish the groceries.

SOLUTION: Remaining number of family members = (30 - 5) = 25

Remaining number of days = (9 - 2) days = 7 days.

30 family members had food for 7 days

1 family member had food for (30 × 7) days [less family member, more days]

25 family members had food for (30 × 7)/25 days [more men, more days] = 8.4 days.

Hence, the remaining food will last for approximately 8 days.

EXAMPLE 18: Painting a house takes three painters 12 hours. How much time will it take nine painters to do the same task?

SOLUTION: First, we need to determine which kind of proportion it is.

It takes 12 hours for 3 painters. So, the 9 painters will take more or less time to do the job?

More painters mean that it'll take less time for them to finish the job. So, we're dealing with inverse proportions.

By using the ***Unitary method***

3 painters are required to finish the painting in 12 hours.

1 painter is required to finish the painting (12× 3) hours. [less painters, more hours]

9 painters are needed to complete the painting (12× 3)/ 9 hours. [more painters, less hours] = 4 hours

Hence, 9 painters will take 4 hours to do the same job.

Level 1: Exercise 2.4 Practice Questions

1. A recipe for 6 people has 5 eggs, 8 potatoes, and 1 cup of cheese.
   a. If the recipe is to be made for 36 people, calculate the quantity required for each of these ingredients.
   b. If the recipe is to be made with 15 eggs, how much of each ingredient is needed, and how many will it serve?

c. If there are 64 potatoes for the recipe, how much of each ingredient is needed, and how many will it serve?

2. For every 2 hours of work, Sameer gets $10.
   a. If Sameer does 10 hours of work, how much money will he get?
   b. If Sameer got $90, how many hours does he work?
   c. If Sameer does 1-hour work, how much will he get?

3. In a basketball game, Kevin finds that for every 8 shots he misses 3 and gets 5 baskets.
   a. If he had 40 shots in a game, how many would he get in at this rate?
   b. How many did he miss if he got 24 shots in 1 game?
   c. How many would he have gotten in if he missed 9 shots in another game?

4. Fertilizer will be mixed by adding 3 tablespoons of fertilizer concentrate to 9 liters of water.
   a. How much concentrate would be used for 18 liters of water?
   b. How much water is needed if 12 tablespoons of fertilizer concentrate are used?
   c. If 3 liters of water is used, how much fertilizer concentrate should be used?

5. Solid plaster comprises 1 part weight of cement, 1 part lime, and 6 parts of sand.
   a. If 6 kg of cement is available, how much lime and sand would need to be added to make the solid plaster?
   b. If 160 kg of plaster are to be made, how much cement, lime, and sand are needed?
   c. If 12 buckets of sand are available, how much cement and lime would be needed to make it into plaster?

6. The ratio of the number of cups of sugar to the number of cups of flour needed to bake cookies is 1:5. If Aisha uses 25 cups of flour, how many cups of sugar are needed?

7. About 13 out of 20 students have smartphones in class. With this ratio, how many students are expected to have a smartphone in the class of 60 students?
8. Sana spent BHD 150 on 3 dresses. How many such dresses can Sana buy for BHD 300?
9. If 9 people consume 3 pizzas, how many people will consume 9 pizzas?
10. A map has a scale of 1cm: 40,000 km. If two cities appear to be 5 cm apart on a map, what is their distance?
11. The price of 32 computers is BHD 2000. Find the number of computers that can be purchased for the BHD 5500.
12. In 5 hours, Sameer drove his car for 150 km. Find the time it takes for him to travel 180 km.
13. It took 2000 workers 4 years to build the mall in Bahrain. How long would it have taken for 3000 workers to build a similar mall?
14. A fully charged I- phone's battery lasts for 48 hours. The duration that the battery lasts is proportional to its percentage charge. How long will the battery last if it is 25% charged?
15. 14 painters can paint a fence in 8 hours.
    a. Find the number of painters needed to complete the job in 2 hours.
    b. How long will it take 16 people to paint it?

Level 2: Exercise 2.5 Practice Questions

1. Four friends want to go hiking. The water they have is enough to last 3 days. 2 more friends join them before they leave and calculate how long will their water last now.
2. A group of 4 students needs 24 days to complete the project. To accomplish a similar task in 16 days, determine how many students are required in a group.
3. A car covers a particular distance in 4 hours with an average speed of 75 km per hour. If the speed increases by 25 km per hour, find the time taken by the car to cover the same distance.

4. If Adam can get BHD 750 profit in a year, at the same rate, find how much profit he can get in 4 years.
5. Sarah takes 15 days to lose 2 kg of weight by doing 30 minutes of exercise per day. If she does exercise for 1 hour and 30 minutes per day at the same rate, how many days will she take to reduce the same weight?
6. Sameer was planning a trip to the US. Before going, he researched and learned that the exchange rate is 2.65 USA for 1 BHD. How much USD would he get if he exchanged 2800 BHD?
7. A cake recipe needs 1.5 cups of milk and 3 cups of flour. Aisha made a mistake and used 5 cups of flour. Find the number of cups of milk she should use to correct the proportions.
8. Adam has an offer on burgers for sale at 20 meals for BHD 17, and Burger King also has an offer for 36 meals for BHD 34. Which one is the better buy?
9. The currency in Egypt is the pound. The exchange rate is approximately 45 Egyptian pounds for every 1 BHD. At this rate, how many BHD would you get if you exchanged 1250 Egyptian pounds?
10. A giraffe is 32 feet tall. A model of it was built with a scale of 1 inch:4 feet. How tall is the model?
11. Suppose the shadow of an 8-foot-tall adult elephant is 4 feet long. Find the height of a baby elephant whose shadow is 1.5 feet long.

2.5 Activity/Experiment—Ratio

Problem 1

- **Step 1**: Draw 7 squares of different side lengths.
- **Step 2**: Measure the length of the diagonal and measure the length of the side of each square in cm.
- **Step 3**: Find the simplest form of the side length ratio to the diagonal length.
- **Step 4**: Write the three-equivalent ratio.

Experiment	Length of the diagonal/ cm	Length of the side/cm	Ratio	Simplest form	Equivalent ratio
Square1					
Square2					
Square3					
Square4					
Square5					
Square6					
Square7					

Problem 2

- **Step 1**- Draw 7 circles of different diameters.
- **Step 2**- Measure the diameter length and calculate the circumferences of the following circles in cm.
- **Step 3**- Find the ratio of the length of the circumference to the length of the diameter. in simplest form
- **Step 4**- Write the three-equivalent ratio.

Experiment	Length of the diagonal/ cm	Length of the side/cm	Ratio	Simplest form	Equivalent ratio
Circle1					
Circle2					
Circle3					
Circle4					
Circle5					
Circle6					
Circle7					

2.6 Activity/Experiment—Proportions

Solving Proportions Maze A!

Directions: Find the value of x for each problem along the way. Use your solutions to navigate through the puzzle, **SHOW ALL STEPS** by circling your answers.

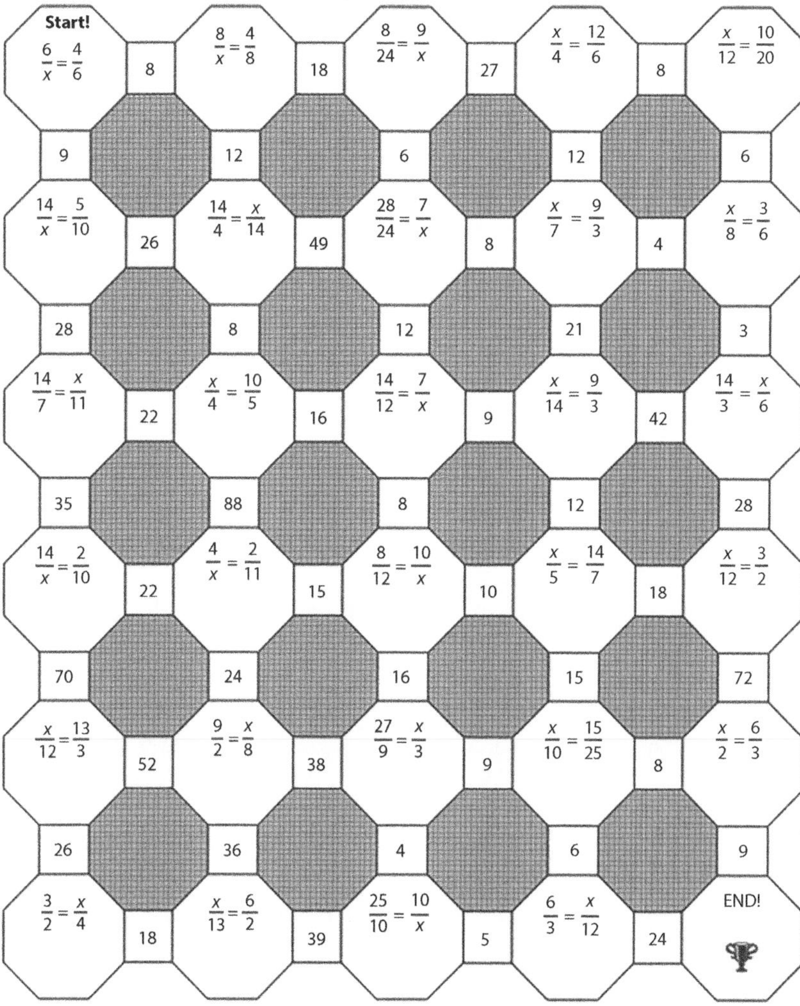

3

Two-Dimensional Geometry

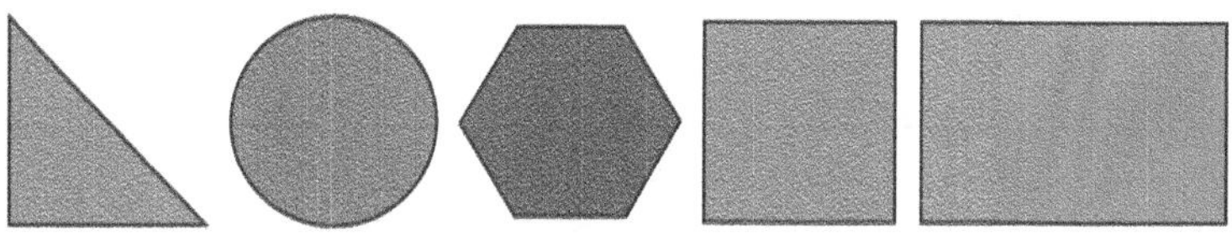

Objectives: This chapter

- *Defines geometry.*
- *Describes geometric figures.*
- *Explain two-dimensional and three-dimensional figures.*
- *Defines the Perimeter and Area of 2D (two-dimensional) shapes.*
- *Defines the surface area and volume of 3D(three-dimensional) shapes.*

3.1 Introduction

Geometry is one of the oldest and most practical branches of mathematics that deals with the shapes and sizes of different shapes and their properties. "GEOMETRY" was derived from a Greek word meaning "EARTH MEASUREMENT."

Objects that have boundary lines, angles, and surfaces are considered SHAPES. A regular is a symmetrical shape such as a square, circle, etc., whereas Irregular shapes are also called freeform shapes or organic shapes; they are asymmetrical. For example: the shape of a tree is irregular or organic.

Ritu Shrivastava, Ramakant Bhardwaj and Satyendra Narayan. Mathematics for Engineers, (59–120)

3.2 Geometry

3.2.1 One-Dimensional Geometry

Point and line: The location of any object with no length, width, or depth is known as a point in. A dot in the figure shows a point. A point is considered zero-dimensional, while a line is one-dimensional, and its length can be measured.

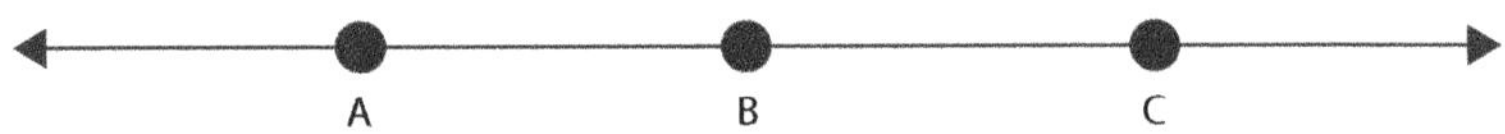

3.2.2 Two-Dimensional Geometry

A flat plane figure or a shape with two dimensions is called a two-dimensional shape. It can be defined as length and width. These shapes are flat and can only be drawn on paper. 2D shapes have different numbers of sides. examples of two-dimensional shapes are a circle, oval (or ellipse), triangle, square, rectangle, and pentagon. Some of these are curved shapes; others comprise only line segments. They are sometimes called plane shapes. There is no thickness in any Two-dimensional or 2-D shapes. They can be measured in only two dimentions.

3.3 Area and Perimeter of 2D Shapes

Area and perimeter in 2D shapes are fundamental terms when dealing with mathematics. It applies to any shape of the field, irregular or regular.

The perimeter is the pathway or the boundary that surrounds a shape. It can also be defined as the length of the outer boundary of a shape.

For example: a house has a fenced yard. The perimeter is the length of the fence. If the yard is 20 m × 20 m, the fence is 20+20+20+20 = 80 m long.

The area is a measure of the surface contained within an object. In the same yard, an area is a measure you need to put grass on the lawn. The area has units of distance squared. For a 20 m × 20 m, you need 20 × 20 = 400 square m of grass.

We can use a rope to measure any small irregular shapes or threads and place it exactly along the shape's boundary once. Finally, the length of the rope along the boundary is the shape's perimeter.

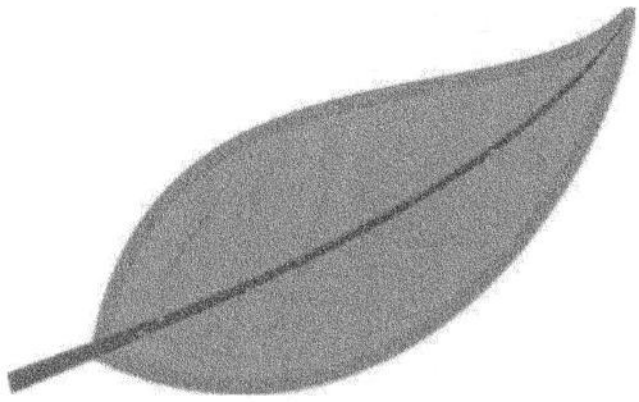

To find the perimeter of this leaf, we will use a thread. Putting thread all around the figure, we can easily find the perimeter of this irregular leaf shape.

3.3.1 Perimeter

We may use the formulas to find its perimeter for any type of polygon – regular or irregular. To better understand the perimeter of a shape in front of you, imagine that we are tiny and standing on one of the shape's corners if we are walking along the shape's sides until we get back to the starting point. The distance we have walked will be the perimeter.

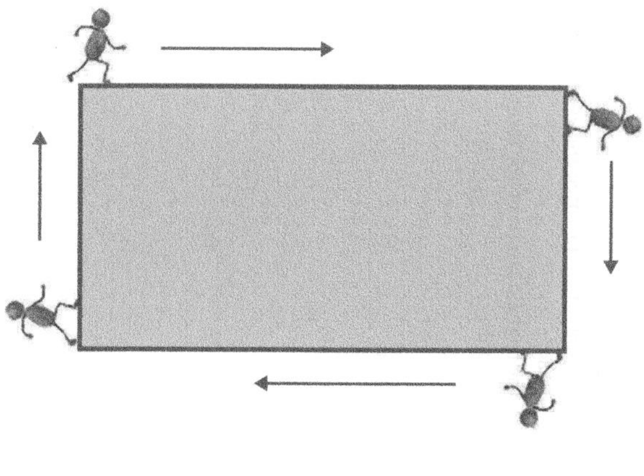

Shapes	Description	Rule of perimeter	Rule of area
Triangles	Any three-sided polygon (2d shape with three edges and three vertices are known as triangles. The sum of all the three angles of a triangle = 180°. The triangle can be of any form, right, acute, or obtuse. 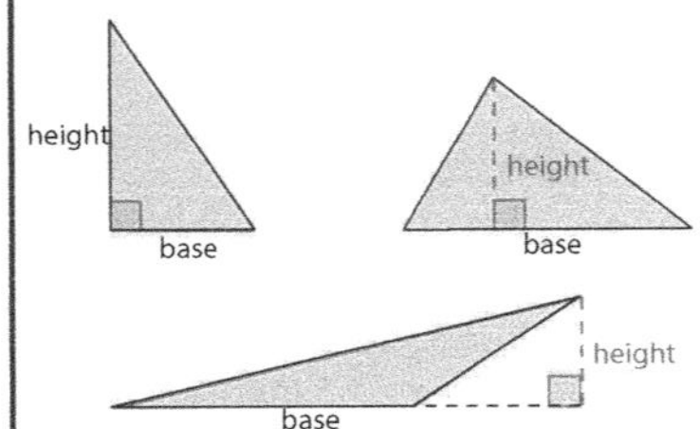	The perimeter of a triangle = the sum of all the three sides = a + b + c.	The area of the triangle $= \frac{1}{2} \times base \times height = \frac{1}{2} \times b \times h$ the Area of the equilateral triangle $= \frac{\sqrt{3}}{4} \times s^2$ since the height of the equilateral triangle is $\frac{\sqrt{3s}}{2}$ THE AREA OF THE SCALENE TRIANGLE CAN BE DETERMINED IN TWO WAYS If the base and height of a triangle are given Area $= \frac{1}{2} \times b \times h$ IF THE SIDES OF A TRIANGLE ARE KNOWN USE HERONS FORMULA Area $= \sqrt{s(s-a)(s-b)(s-c)}$ Area $= \sqrt{[s-(side\, a)][s-(side\, b)][s-(side\, c)]}$ Where a, b, and c are considered as the sides. and s is the semi-perimeter, $s = \frac{a+b+c}{2}$

(Continued)

(*Continued*)

Shapes	Description	Rule of perimeter	Rule of area
Scalene Triangle	Any triangle with all three sides having different lengths is called Scalene Triangles. In this case, all three angles are of different magnitude/measure. 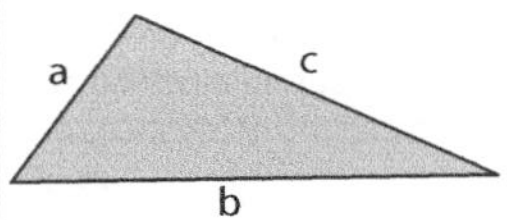		
Isosceles Triangle	Any triangle in which two of the three sides are equal is called an Isosceles Triangle. In this case, the two base angles of the triangle are of equal magnitude/measure. 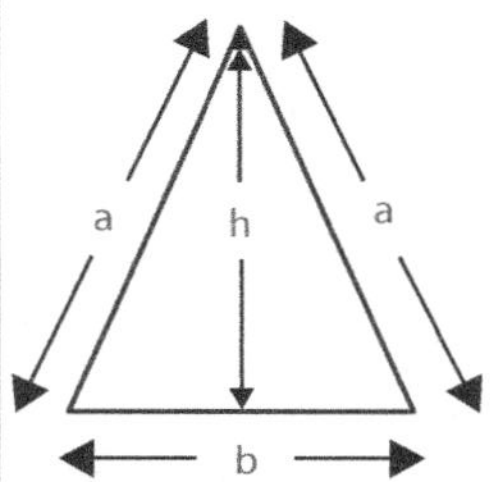		

(*Continued*)

(*Continued*)

Shapes	**Description**	**Rule of perimeter**	**Rule of area**
Equilateral Triangle	Any triangle in which all three sides of the triangles are equal is called an equilateral triangle. In this case, all three angles are of the same magnitude/measure (60°) 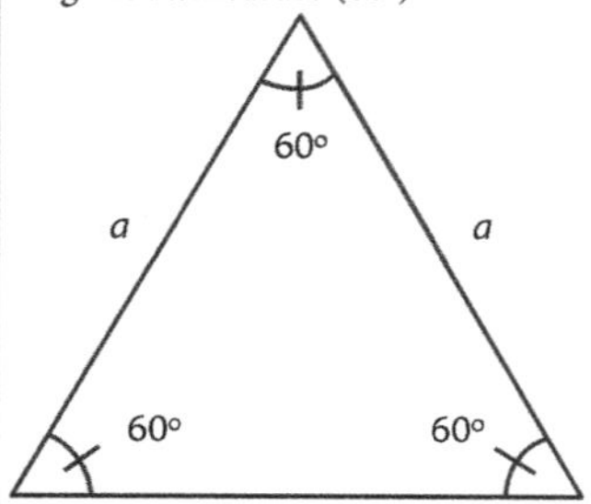		
Quadrilateral	The quadrilateral is a simple closed figure (polygon) that has four-line segments. 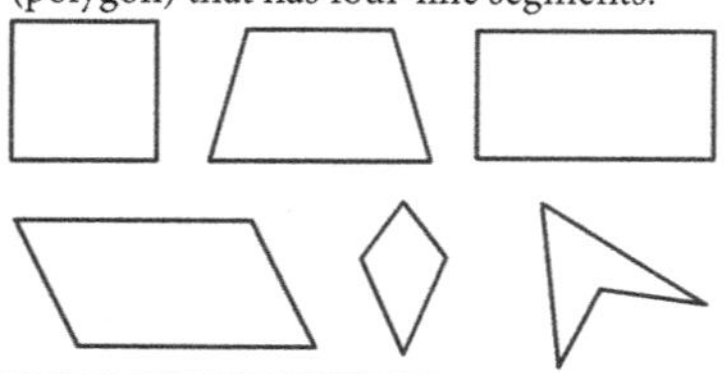		

(*Continued*)

(Continued)

Shapes	Description	Rule of perimeter	Rule of area
Square	Square: A square is a four-sided polygon; it is considered a two-dimensional regular quadrilateral (2D shape) with equal four sides and all the angles. Each angle of the square is 90°, and its diagonals also bisect each other at 90°. A wall or a table with equal sides are EXAMPLEs of square shapes. 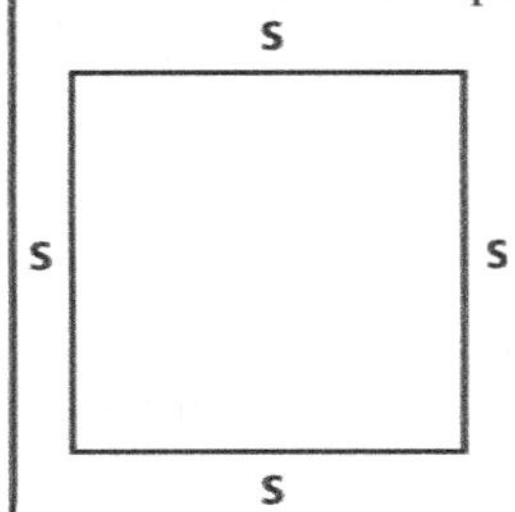	The perimeter of the square = 4 x side of the square.	Area of square = product of base and height For any square, the base(a) is the same as the height(a) $= a \times a = (side)^2$

(Continued)

(*Continued*)

Shapes	Description	Rule of perimeter	Rule of area
Rectangle	A rectangle is a 2D shape with four sides, where the opposite sides are equal and parallel. All the angles of a rectangle are equal to 90°. Brick, TV, cardboard, etc., are EXAMPLEs of rectangles. 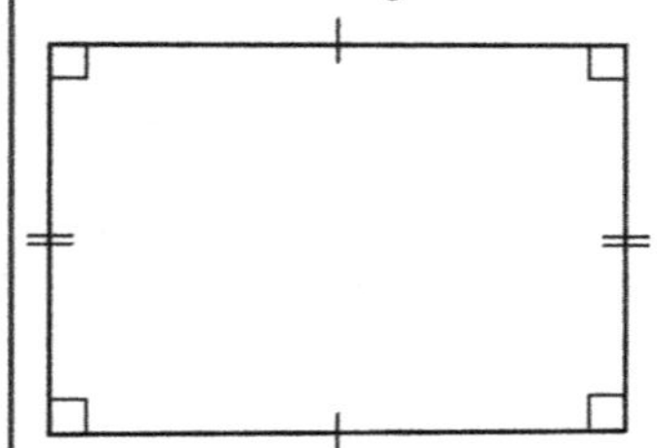	The perimeter of a rectangle = Perimeter of rectangle = 2 (length + width) $= 2(l+b)$	Area of rectangle = product of two different sides of the rectangle. These two sides can be base or height, length or width or breadth. $= l \times h = l \times b = l \text{ x } w = h \times w$

(*Continued*)

(Continued)

Shapes	Description	Rule of perimeter	Rule of area
Parallelogram	A parallelogram is a special type of quadrilateral that has equal and parallel opposite sides. In the figure, ABCD is a parallelogram, in which AB. is parallel to CD and AD is parallel to BC. Also, AD = BC and AB = CD. A parallelogram is a slanting rectangle with two sets of parallel opposite sides. 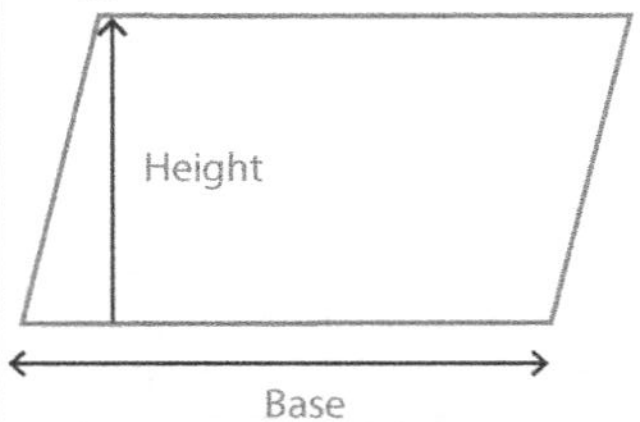	Perimeter of parallelogram = 2x sum of adjacent sides =2 (length + width) = 2(*l*+*b*)	Area of parallelogram= base x height Or can also be learned as Area = side × perpendicular on the side = b × h
Trapezium	A trapezium, also known as a trapezoid, is a 2d shape and a type of quadrilateral, which has only two parallel sides, and the other two are non-parallel. 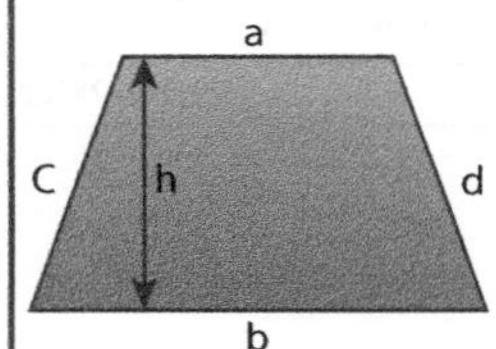	The perimeter of Trapezium, P = a + b+ c + d units	Area of Trapezium $= 1/2 \times h \times (a+b)$ = 1/2 (sum of parallel sides) × (perpendicular distance between them)

(Continued)

(Continued)

Shapes	Description	Rule of perimeter	Rule of area
Rhombus	Rhombus refers to a four-sided (2D shape) figure with congruent sides. The opposite angles of a rhombus and four sides are equal. Its diagonals meet at 90°, therefore, perpendicular to each other. Its adjacent angles are supplementary, which means that the sum of their measure is equal to 180°. It is also known as an equilateral parallelogram. 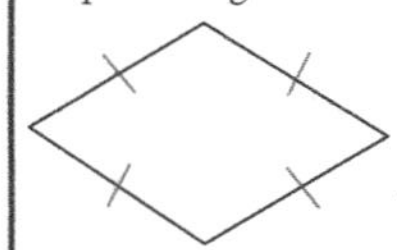 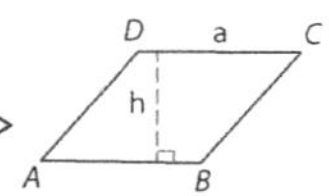 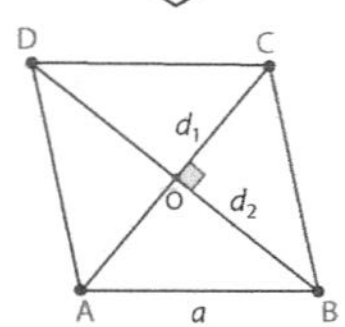	Perimeter of Rhombus = 4 x Side of the Rhombus.	Area of Rhombus = base × height = side x perpendicular on the side = a × h Area of Rhombus = $(d_1 \times d_2)/2$ where d_1 and d_2 are the diagonals of the Rhombus.

Conversion formulas of the UNITS OF AREA

Length units	Area units
1 m = 100 cm	$1\ m^2 = (100 \times 100)\ cm^2 = 10000\ cm^2$
1 m = 10 dm	$1\ m^2 = (10 \times 10)\ dm^2 = 100\ dm$
1 cm = 10 mm	$1\ cm^2 = (10 \times 10)\ mm^2 = 100\ mm^2$
1 km = 1000 m	$1\ km^2 = (1000 \times 1000)\ m^2 = 1000000\ m^2$
1 hm = 100 m	$1\ hm^2 = (100 \times 100)\ m^2 = 10000\ m^2 =$ 1 hectare (ha)
1 dam = 10 m	$1\ dam^2 = (10 \times 10)\ m^2 = 100\ m^2 = 1$ acres

Note: hm, hectometer; dam, decameter

3.3.2 Area of Two-Dimensional Geometry

The region enclosed by any 2D shape is known as its area in geometry. The total number of unit squares that cover the surface of the full figure is known as the area of the figure. The units to measure area are square units such as square centimeters, square feet, square inches, etc.

3.3.3 Finding the Area of Shapes by Counting the Square

An easier way to calculate the area of the given figure is to count the number of squares included in the figure. Since all squares are not complete in the figure, so we follow the following rules to find the area of the figure: -

a. If the area is less than half square, ignore it (take it 0).
b. If the area is more than half a square, take 1 square unit.
c. If the area is exactly half square, take it ½ square units.

To find an area value using the grid method, we need to know the size of a grid square. This example uses centimeters, but the same procedure applies to any unit of length or distance. It can be inches, meters, miles, feet, etc.

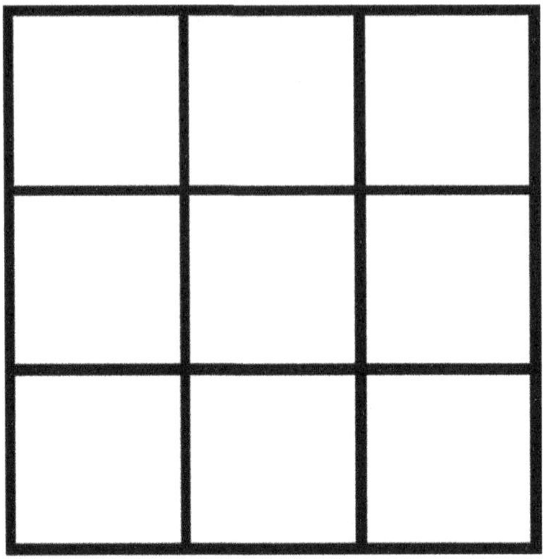

In this example, each grid square has a width of 1cm and a height of 1cm. In other words, each grid square is one 'square centimeter.'

Count the grid squares inside the large square to find its area. There are nine small squares, so the area of the large square is 9 square centimeters. In mathematics, we abbreviate 'square centimeters' to cm^2. The two means 'squared.' Each grid square is 1 cm^2. The area of the big square is 9 cm^2.

For example: Find the Area of the following figures by counting squares:

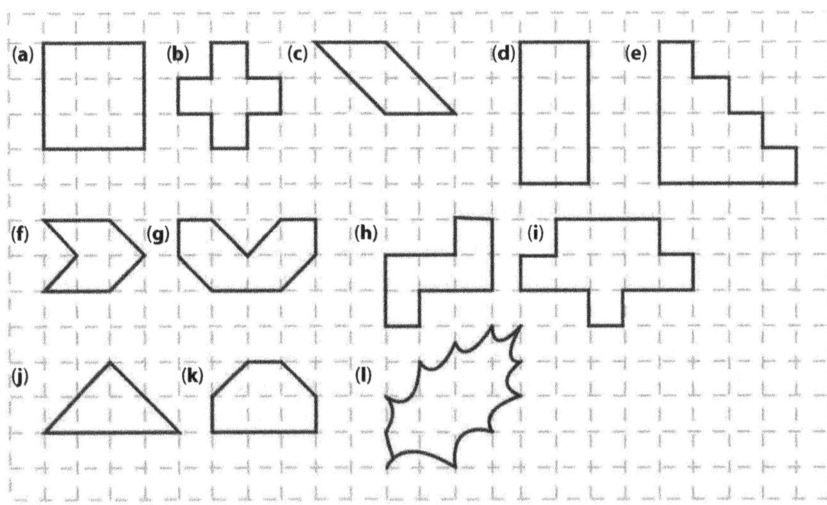

SOLUTION:

(a) There is a total of nine filled squares.
Area covered by 1 square = 1 unit^2
∴ Area covered by 9 squares = 9×1 =9 unit^2
(b) There is a total of 5 filled squares.
Area covered by 1 square = 1 unit^2
∴ Area covered by 5 squares = 5×1 =5 unit^2
(c) There is total fully 2 filled squares and 4 half-filled squares.
Area covered by 1 square = 1 unit^2
∴ Area covered by 2 squares = 2×1 =2 unit^2
And the area covered by half-filled squares $= 4\times\frac{1}{2} = 2$ unit^2
∴ Total area = 2 + 2 = 4 unit^2
(d) There is a total of 8 filled squares.
Area covered by 1 square = 1 unit^2
∴ Area covered by 8 squares = 8×1 =8 unit^2
(e) There is a total of 10 filled squares.
Area covered by 1 square = 1 unit^2
∴ Area covered by 10 squares = 10×1 =10 unit^2
(f) There is total fully 2 filled squares and 4 half-filled squares.
Area covered by 1 square = 1 unit^2
∴ Area covered by 2 squares = 2×1 =2 unit^2
And the area covered by half-filled squares $= 4\times\frac{1}{2} = 2$ unit^2
∴ Total area = 2 + 2 = 4 unit^2
(g) There is total fully 4 filled squares and 2 half-filled squares.
Area covered by 1 square = 1 unit^2
∴ Area covered by 4 squares = 4×1 =4 unit^2
And the area covered by half-filled squares $= 2\times\frac{1}{2} = 1$ unit^2
∴ Total area = 4 + 1 = 5 unit^2
(h) There are a total of 5 filled squares.
Area covered by 1 square = 1 unit^2
∴ Area covered by 5 squares = 5×1 =5 unit^2
(i) There is a total of 9 filled squares.
Area covered by 1 square = 1 unit^2
∴ Area covered by 9 squares = 9×1 =9 unit^2
(j) There is total fully 2 filled squares and 4 half-filled squares.
Area covered by 1 square = 1 unit^2
∴ Area covered by 2 squares = 2×1 =2 unit^2
And the area covered by half-filled squares $= 4\times\frac{1}{2} = 2$ unit^2
∴ Total area = 2 + 2 = 4 unit^2
(k) There is total fully 4 filled squares and 2 half-filled squares.

Area covered by 1 square = 1 unit^2
∴ Area covered by 2 squares = 4×1 =4 unit^2
And the area covered by half-filled squares $= 2 \times \frac{1}{2} = 1$ unit^2
∴ Total area = 4 + 1 = 5 unit^2
(l) There is total fully 3 filled squares and 10 half-filled squares.
Area covered by 1 square = 1 unit^2
∴ Area covered by 3 squares = 3×1 =3 unit^2
And the area covered by half - filled squares $= 10 \times \frac{1}{2} = 5$ unit^2
∴ Total area = 3 + 5 = 8 unit^2

NOTE: - This method of calculation is not very accurate, so we find the area of 2D figures using formulas

EXAMPLE 1: Find the Perimeter of the following triangles

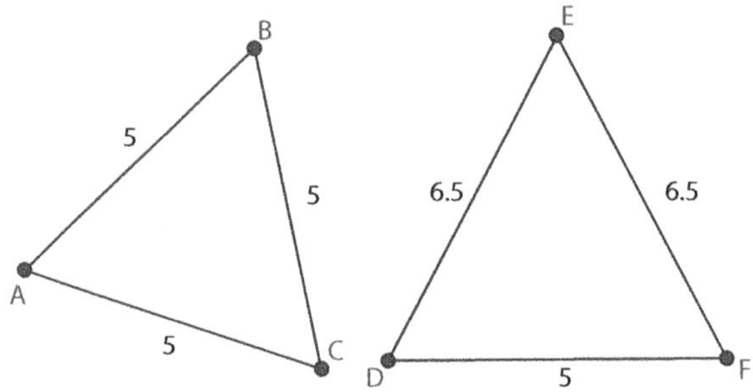

SOLUTION: The Perimeter of the triangle is the sum of the length of all the sides

Perimeter of Triangle ABC = 5 + 5 + 5 = 15 units
Perimeter of Triangle DEF = 6.5 + 6.5 + 5 = 18 units

EXAMPLE 2: If the perimeter of an equilateral triangle (all three sides equal) is 84 cm. Find its area and the length of each side. (The height of the equilateral triangle is $\frac{\sqrt{3}}{2} \times s$)

SOLUTION: Perimeter of the triangle is the sum of all the sides

For an equilateral triangle, all the sides are equal; let "a" be the length of each side of the triangle

$$\text{Hence,} \qquad a + a + a = 84$$

$$3a = 84$$

$$a = \frac{84}{3} = 28 \text{ cm.}$$

Therefore, each side of the triangle is 28 cm.

$$\text{Area} = \frac{1}{2} \times b \times h$$

$$= \frac{1}{2} \times s \times \frac{\sqrt{3}}{2} \times s$$

$$= \frac{\sqrt{3}}{4} \times s^2 = \frac{\sqrt{3}}{4} \times 28^2 = 678.96 \text{ unit}^2$$

EXAMPLE 3: The ratio of the sides of a scalene triangle (all sides are unequal) is 4:2:3; if its perimeter is 27 inches, find the lengths of the sides of the triangle. Also, find its area.

SOLUTION: Let $4x$, $2x$, and $3x$ be the sides of the triangle.

Perimeter = 27 inches

So, $2x + 4x + 3x = 27$

$9x = 27$

Dividing by nine into both sides, $x = 3$

The sides of the triangle are 2×3, 4×3 and 3×3

Therefore, the sides of the triangle are 6 inches, 12 inches, and 9 inches.

As sides are known, the area can be determined by **HERONS FORMULA**

HERONS FORMULA: a formula used to find the area of the triangle when all three sides are given.

$$\text{Area} = \sqrt{s(s-a)(s-b)(s-c)}$$

$$\text{semi-perimeter, } s = \frac{a+b+c}{2} = \frac{6+12+9}{2} = \frac{27}{2} = 13.5$$

$$\text{Area} = \sqrt{\left[s\left(s - sidea\right)\right]\left[s-\left(sideb\right)\right]\left[s-\left(sidec\right)\right]}$$

$$= \sqrt{13.5[13.5-6][13.5-12][13.5-9]}$$

$$= \sqrt{[13.5][7.5][1.5][4.5]} = 683.438 \text{ inch}^2$$

EXAMPLE 4: If the length of the diagonals of a square is $2\sqrt{2}$ cm. Find its perimeter and area.

SOLUTION: To calculate the perimeter of a shape, we first need to know its side length.

Let 's' be the length of each side of the square.

In the figure shown below, consider the right triangle ABC.

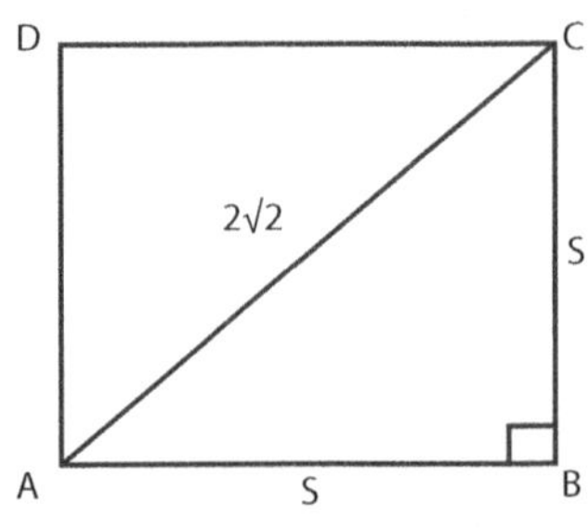

By using the Pythagorean Theorem,

$AB^2 + BC^2 = AC^2$

$S^2 + S^2 = 2 \times (\sqrt{2})^2$ (by Substituting values)

$2S^2 = 2 \times 2 \times (\sqrt{2})^2$

$2S^2 = 4 \times (2)$ (Simplify and solve for S)

$2S^2 = 8$

$S^2 = 4$ (by dividing each side by 2)

$S = 2$ (positive square root on both sides)

Perimeter $= 4S = 4 \times 2 = 8$ cm

Therefore, the perimeter of the square is 8 cm. and the side of the square = 2 cm

Area of Square = $(\text{side})^2$

$= 2^2 = 4 \text{ cm}^2$

EXAMPLE 5: Calculate the Perimeter and the Area of each shape by counting the squares (use 1 small square = 1 in)

1)

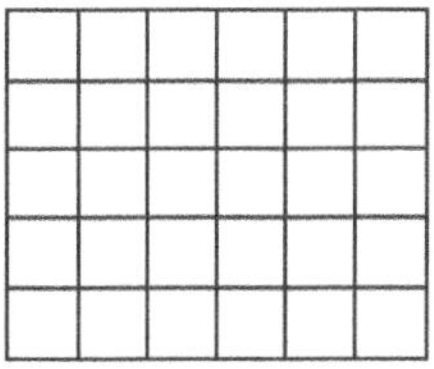

2)

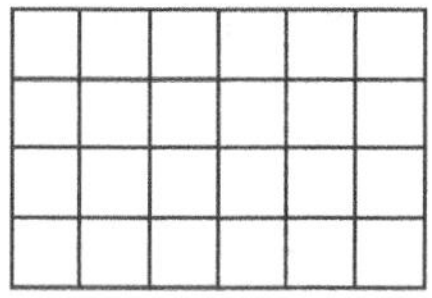

3)

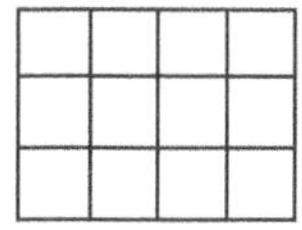

SOLUTION: Perimeter of the rectangle is

1) Perimeter = 6 + 5 + 6 + 5 = 22 in

$$\text{Area} = 5 \times 6 = 30 \text{ in}^2$$

2) Perimeter = 6 + 4 + 6 + 4 = 20 in

$$\text{Area} = 6 \times 4 = 24 \text{ in}^2$$

3) Perimeter = 4 + 3 + 4 + 3 = 14 in

$$\text{Area} = 4 \times 3 = 12 \text{ in}^2$$

However, we can use a formula to calculate the perimeter of the rectangle given as follows: -

Perimeter of rectangle = 2 × (length + width)

EXAMPLE 6: A park is 150 m long and 350 m wide. Find the cost of fencing it at $ 5.50 per meter.

SOLUTION: Length of the park =150 m
Width of the park =350 m
Fencing is only on the boundary of the park is the Perimeter.
Perimeter = 2(length + width)
Perimeter = 2(150 + 350) = 2 × 500 = 1000 m
Cost of fencing = perimeter × per unit cost = 1000 × 5.5 = $ 5500
Cost of fencing the park = $ 5500

EXAMPLE 7: The sides of a rectangle are in the ratio 4: 5. If its perimeter is 720 m, find its length and width.

SOLUTION: Sides of the rectangular are given to be the ratio of 4:5

Let the sides of the rectangle be $4x$ and $5x$

Perimeter = 2(L+W)

$$720 = 2(4x + 5x)$$

$$720 = 2(9x) = 18x$$

$$x = \frac{720}{18} = 40$$

The sides of the rectangle are 4×40 and 5×50 =160 cm and 200 cm.

EXAMPLE 8: A wooden tabletop is of width 6 m and perimeter 20 m. Find the length and the area of the tabletop.

SOLUTION: Width of the tabletop, W = 6m

The perimeter of the tabletop (P) = 20m

Let the length of the tabletop = L

Perimeter of rectangle P=2(L+W)

$$2(L+W) = 20$$

$$2\ (L + 6) = 20$$

Dividing by 2 into both sides,

$$6 + L = 10$$

$$L = 4$$

Therefore, the length of the wooden cupboard is 4 m.

$$\text{Area} = L + W = 4 \times 6 = 24\ m^2$$

EXAMPLE 9: A rectangular cable wire has 40 cm × 20 cm dimensions. Calculate and find its side if it is bent as a square. Will the Area of the Rectangle and the square be equal?

SOLUTION: When the wire is bent from rectangle to square, the perimeter of the two shapes will remain the same; only the shape gets changed.

Therefore, the Perimeter of the rectangle = Perimeter of the square

2(L+W) = 4 × side of the square

2 (40+20) = 4 × side of the square

Side of the square = 2(40+20)/ 4 = 120/4 = 30 cm

Area of the Square = $(\text{side})^2 = (30)^2 = 900\text{cm}^2$

Area of the Rectangle = L + W = 40 × 20 = 800 cm^2

Hence the Area of the Rectangle and the square will not be equal.

EXAMPLE 10: The dimensions of a park are 4 m × 3 m. Find the length of a pipe to be fixed diagonally in the park.

SOLUTION: The tallest pole that can be placed in the park will be along the diagonal of the park.

Using Pythagoras' theorem, we get

$$Diagonal^2 = Lenght^2 + width^2$$

$$D^2 = 3^2 + 4^2$$

$$D^2 = 9 + 16$$

$$D^2 = 25$$

$$D = 5$$

So, the length of the pipe that is to be placed in the park = 50 m

EXAMPLE 11: The diagonal of a square ABCD is 32 cm. Calculate its Perimeter and Area. (Take $\sqrt{2}$=1.41)

SOLUTION: Given that the diagonal of the square= 32 cm

To calculate the area and the perimeter, we need to find the side of the square.

Using Pythagoras' theorem, we get

$$D^2 = side^2 + side^2$$

$$(32)^2 = 2 \times side^2$$

$$32 = \sqrt{2} \times side$$

$$\text{side} = \frac{32}{\sqrt{2}} = 16\sqrt{2}$$

$$\text{Perimeters of the square} = 4 \times side$$

$$\text{Perimeters of the square} = 4 \times 16\sqrt{2}$$

$$\text{Perimeters of the square} = 64\sqrt{2}$$

$$\text{Perimeters of the square} = 64 \times 1.41$$

$$\text{Perimeters of the square} = 90.24 \text{ cm}$$

$$\text{Area of the square} = side^2 = (16\sqrt{2})^2 = 512 \text{ cm}^2$$

EXAMPLE 12: A rectangular garden has a length of 120 m and a width of 90 m. Find the number of trees that can be planted at 7m spacing between each tree around the garden.

SOLUTION: First, we find the perimeter of the rectangle garden
Perimeter =2 (L + W) = 2(120 + 90) = 2(210) = 420 m

A tree is planted at 7 meters; therefore, the number of trees planted =

$$\frac{Perimeter}{7} = \frac{420}{7} = 60 \text{ trees}$$

EXAMPLE 13: The length of a rectangle is three times its width. If the length of the diagonal is 81 cm, then what is the perimeter of the rectangle? (Take $\sqrt{10} = 3.16$

SOLUTION: Let the width of the rectangle = x, Length of the rectangle = $3x$ $Diagonal^2 = length^2 + width^2$

$$81^2 = (3x)^2 + (x)^2$$

$$81^2 = 9x^2 + x^2$$

$$x = \frac{81}{\sqrt{10}} = \frac{81\sqrt{10}}{10}$$

$$= 8.1 \times 3.16 = 25.596$$

Width of rectangle = 25.596 m
Length of rectangle = 76.788 m
Perimeter of rectangle = 2(L+W)
Perimeter of rectangle = 2(25.596 +76.788) = 204.768 m

EXAMPLE 14: The lengths of the adjacent sides of a parallelogram are (x + 5) yards and (–2 + 3x) yards. Determine the value of x if its perimeter is 90 yards.

SOLUTION: Perimeter of parallelogram =2 × (sum of adjacent sides)

$$90 = 2 \times (x + 5 - 2 + 3x)$$

$$90 = 2 \times (4x + 3)$$

$$90 = 8x + 6$$

$$8x = 84$$

$$x = 10.5$$

EXAMPLE 15: Calculate the Perimeter of this Rhombus:

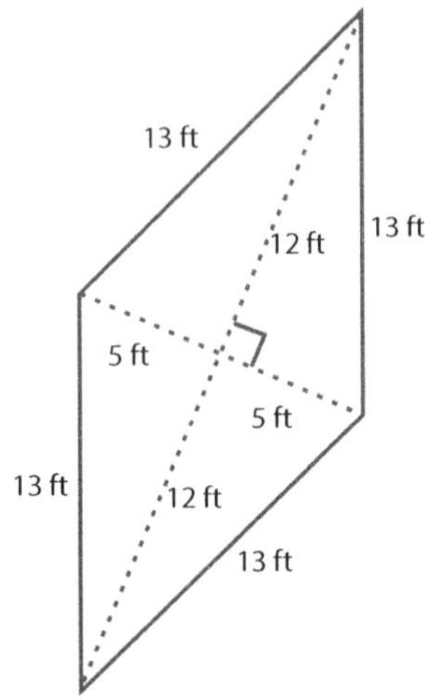

SOLUTION: Perimeter of Rhombus = 4 × side = 4 ×13 = 52 ft

EXAMPLE 16: The Perimeter of a regular pentagon is 40 cm. Find the length of the side.

SOLUTION: A pentagon is five figures. In a regular pentagon, all the sides are equal

$$\text{The perimeter of the pentagon} = 5 \times \text{side}$$

$$40 = 5 \times \text{side}$$

$$\text{side} = \frac{40}{5} = 8 \text{ m}$$

Each side of the pentagon is 8 m.

Fact: The Pentagon building, which is the headquarters of the US Department of Defence, is an excellent example of the pentagon shape

EXAMPLE 17: Find the perimeter of the following figure.

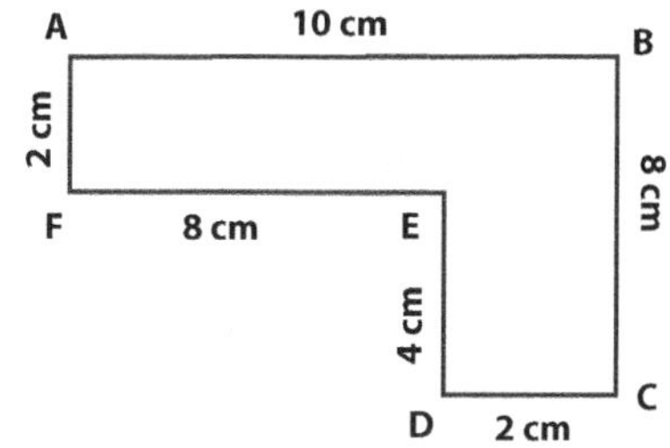

SOLUTION: The perimeter of the figure = sum of all the sides =10 + 8 + 2 + 4 + 8 + 2 = 34cm

EXAMPLE 18: Find the value of x and hence the unknown sides if the perimeter of the given figure is 20 yards.

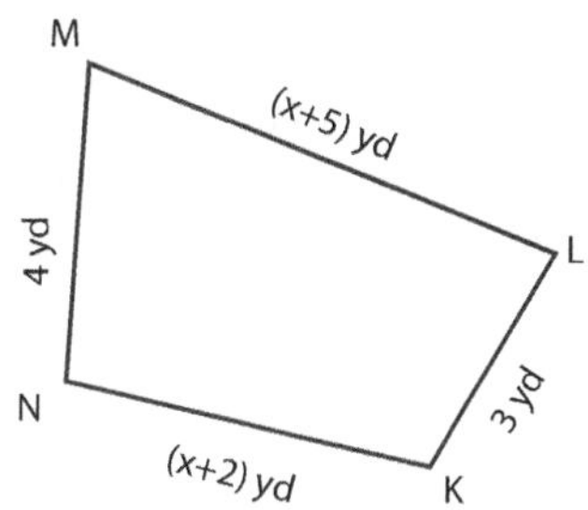

SOLUTION: Perimeter of figure = sum of all sides

$$20 = (x+5)+3+(x+2)+4$$
$$20 = (2x+14)$$
$$2x = 20-14$$
$$2x = 6$$
$$x = 3$$
$$\text{ML} = x+5 = 3+5 = 8 \text{ yd}$$
$$\text{NK} = x+2 = 3+2 = 5 \text{ yd}$$

EXAMPLE 19: A farmer has a field shaped like a regular heptagon with a perimeter of 700 m. What is its side?

SOLUTION: A regular heptagon has seven equal sides

The perimeter of the heptagonal field = 7 × sides

$$\text{Side} = \frac{700}{7} = 100 \text{ m}$$

Therefore, each side of the heptagon in the field is 100 m.

3.4 Curved Plane Figures

3.4.1 Circles

Circles are two-dimensional round-shaped figures formed by tracing all points in a plane such that they are at an equal distance from a fixed point, the center. The constant distance is called its radius. The word circle is derived from the Greek word Kirkos, meaning hoop or ring.

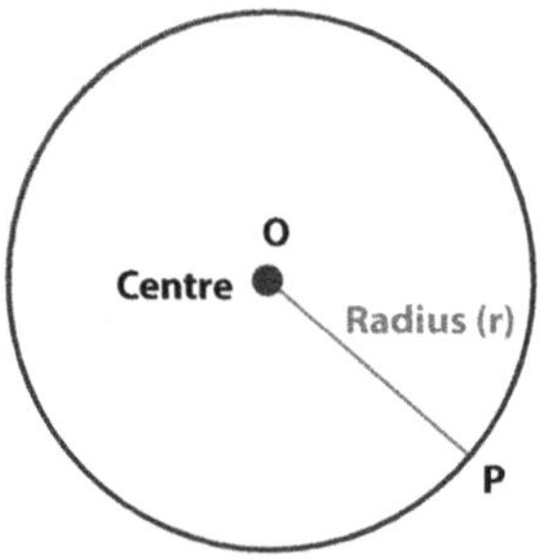

The circumference is the circle's boundary, and its area is a flat surface inside its boundary.

Its circumference is given as $\mathbf{C = 2\pi r}$ where r is the circle's radius. Where 'π' is called **pi** and is taken as $\frac{22}{7}$ or **3.14** (approximately)

ALL ABOUT CIRCLES

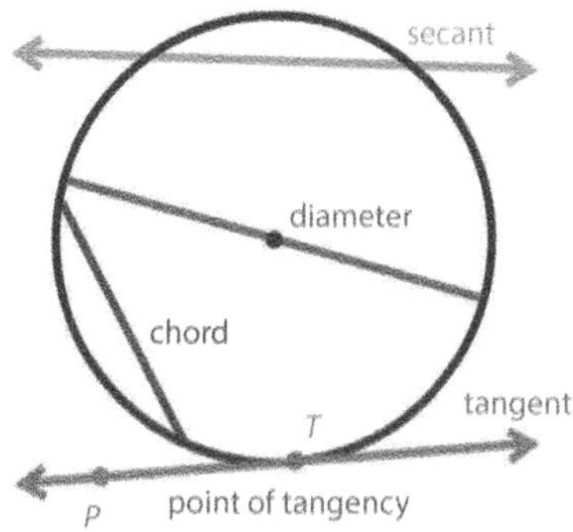

RADIUS	The distance from the center of the circle to its outer rim. It is usually denoted by r
CHORD	A line segment whose endpoints are on a circle
DIAMETER	A chord passes through the center of the circle. It is the longest chord of the circle. The length of the diameter is two times the length of the radius. It is usually denoted by d. And d = 2r.
SECANT	A line that intersects a circle in two points is called a secant
TANGENT	A line that intersects a circle at exactly one point. Or the line which touches a circle is called a tangent.
POINT OF TANGENCY	The point where a tangent line touches the circle is called the point of tangency. Every point on the boundary of a circle can be a point of tangency.

In geometry, the area enclosed by a circle of radius r is πr^2. Here, the Greek letter π represents the constant ratio of the circle's circumference to its diameter, approximately equal to 22/7 or 3.14159.

The area is given as $\mathbf{A = \pi r^2}$

3.4.2 Semi-Circles

A semi-circle is one-half of a circle; it is formed by splitting a circle into two equal parts.

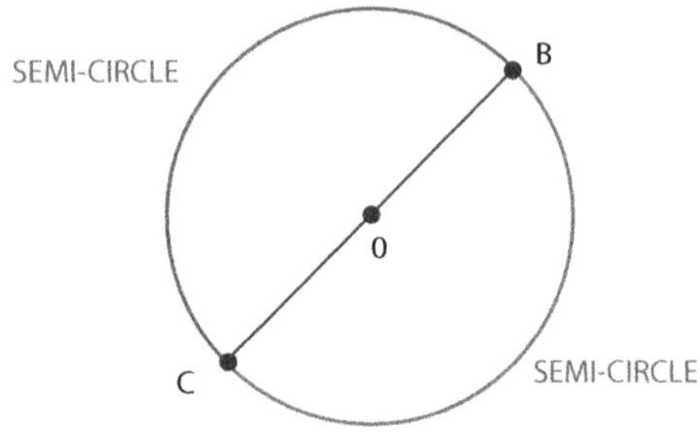

The perimeter (circumference) of a semi-circle is given as $C = \frac{1}{2}(2\pi r) + 2r = \pi r + 2r = r(\pi + 2)$

For the semi-circle, the area is given as $\boxed{\boldsymbol{A} = \frac{\pi \boldsymbol{r}^2}{2}}$

3.4.3 Quarter Circles (QUADRANT)

One-fourth of a circle is known as a quadrant of a circle; it is formed by splitting a circle into four equal parts or a semi-circle into two equal parts.

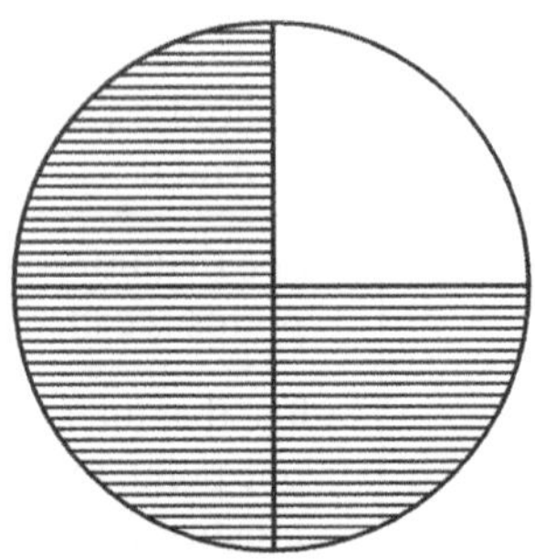

In the given figure the unshaded area is a quarter circle. The perimeter (circumference) of a quadrant is given as $C = \frac{1}{4}(2\pi r) + 2r = \frac{\pi r}{2} + 2r = \frac{\pi r}{2} + d$

(d is the diameter, and r is the circle's radius).

For the quarter circle, the area is given as $\boldsymbol{A} = \frac{\pi \boldsymbol{r}^2}{4}$

Table of Rules

Shape	**Circumference**	**Area**
Circle	$2\pi r$	πr^2
Semi-circle	$A = \pi r + d \text{ or } \pi r + 2r \text{ or } r(\pi + 2)$	$A = \frac{\pi r^2}{2}$
Quarter-circle	$\frac{\pi r}{2} + 2r = \frac{\pi r}{2} + d$	$\frac{\pi r^2}{4}$

EXAMPLE 20: What is the area of this circle?

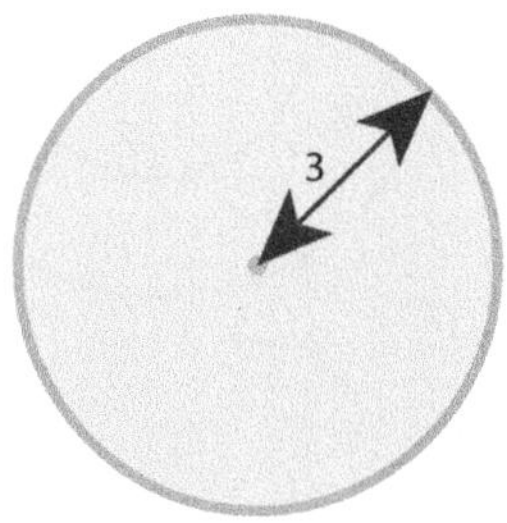

SOLUTION: Radius = r = 3
Area= $\pi \times r^2 = \pi \times 3^2 = \pi \times (3 \times 3) = 3.14159... \times 9 = 28.27$ (to 2 decimal places)

EXAMPLE 21: The diameter of a wheel is 100 cm. How many revolutions will the wheel make to travel 154 meters?

SOLUTION: Diameter of wheel = 100 cm
Radius of wheel = 50 cm

$$\text{Circumference of the wheel} = 2\pi r = 2 \times \frac{22}{7} \times 50 = \frac{2200}{7} \text{ cm}$$

$\frac{2200}{7}$ *cm* distance covered in one revolution.

$$154 \text{ meters} = 15400 \text{ cm distance covered in } = \frac{1}{\frac{2200}{7}} \times 15400 = 49$$

The wheel will make 49 revolutions to travel a distance of 154 m

EXAMPLE 22: A road runs around a circular playground. If the difference in their circumferences

(Circular road and the circular playground) are 33 meters, find the width of the road.

SOLUTION: Let the radius of the outer circle $= r_1$, and the radius of the outer circle $= r_2$

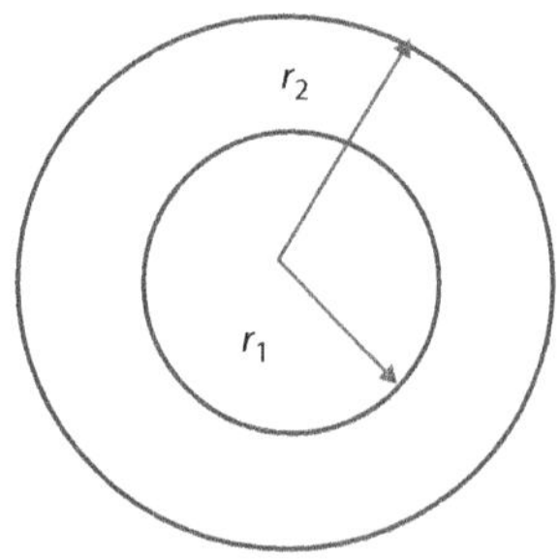

According to the equation

$$2\pi r_2 - 2\pi r_1 = 33$$

$$2\pi (r_2 - r_1) = 33$$

$$r_2 - r_1 = \frac{33}{2\pi} = \frac{33 \times 7}{2 \times 22} = 5.25$$

Hence, the width of the road is 5.25 m.

EXAMPLE 23: If the circumference of a circle is 37m more than the radius. Find the circumference of the circle.

SOLUTION: Ler the circumference is c, and the radius of the circle is r. Given that the difference in circumference and radius of the circle is 37 m.

$$c - r = 37$$
$$2\pi r - r = 37$$
$$r\left(2\times\frac{22}{7}-1\right)=37$$
$$r\left(\frac{44-7}{7}\right)=37$$
$$r\left(\frac{37}{7}\right)=37$$
$$r = 7 \text{ m}$$

Circumference of the circle is $= 2\pi r$

Circumference of the circle is $= 2\times\frac{22}{7}\times 7$

Circumference of the circle is $= 44\ m$

EXAMPLE 24: If the diameter of a semi-circle is 7 cm. Find its circumference.

SOLUTION: $\text{r} = \frac{\text{diameter}}{2} = \frac{7}{2} = 3.5 \text{ cm}$

The circumference of the semicircle $= \pi r + 2r = (\pi + 2)r$

here $\text{r} = 3.5$ and $\pi = \frac{22}{7}$

$= [(\frac{22}{7}) + 2]\times 3.5$

$= (22 + 14)\times 0.5 = 36\times 0.5 = 18 \text{ cm}$

EXAMPLE 25: A square-shaped wire length is 108 cm. The wire is unbent and twisted into the shape of a semi-circle. Calculate the radius of the semi-circle.

SOLUTION: The perimeter of the square = Perimeter of the semi-circle

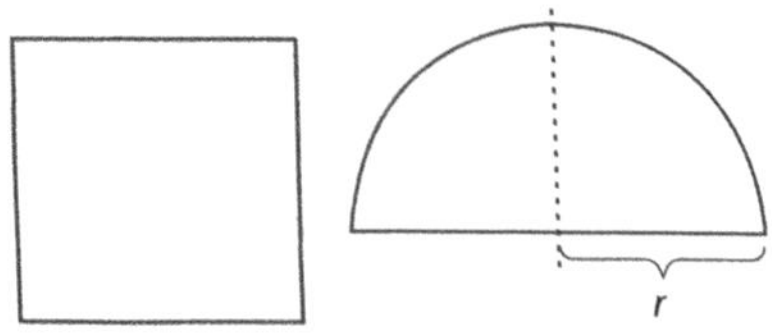

$$4 \times a = \pi r + 2r$$

$$108 = \pi r + 2r = r(\pi + 2)$$

$$108 = r(\frac{22}{7} + 2) = r(\frac{22 + 14}{7})$$

$$r = \frac{108 \times 7}{36} = 21\ cm$$

3.5 Perimeter and Area of Irregular Polygons and Composite Shapes

There are fixed formulas to calculate the perimeter and area of any regular shapes, such as squares, rectangles, and circles. But if the shape is made up of a combination of other shapes (Compound shapes), we must identify and calculate the missing sides by using some given information or sides.

Here are a few steps to follow to find the Perimeter:

1. Divide the combined or irregular shapes into regular shapes by drawing straight line/es
2. Calculate missing perimeter measurements from the given measurement
3. Add all the boundary terms to find the perimeter of the complete figure.

For example: look at the given compound shape below. It is easy to find its perimeter if each side's length is known.

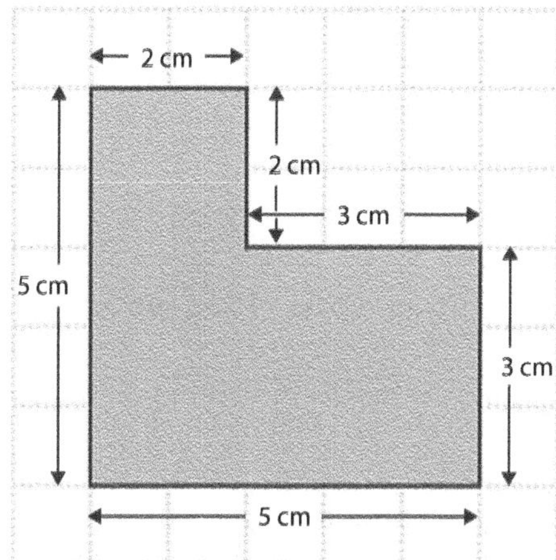

Perimeter =2 cm + 2 cm + 3 cm + 3 cm + 5 cm + 5 cm = 20 cm

Steps to Find Area

1. Draw a few straight lines to identify some regular shapes from a given irregular shape.
2. Calculate the missing side length from the given information.
3. Find the area of each regular shape separately.
4. Add or subtract accordingly.

Note: total area will be the same whether you draw the straight line horizontally or vertically.

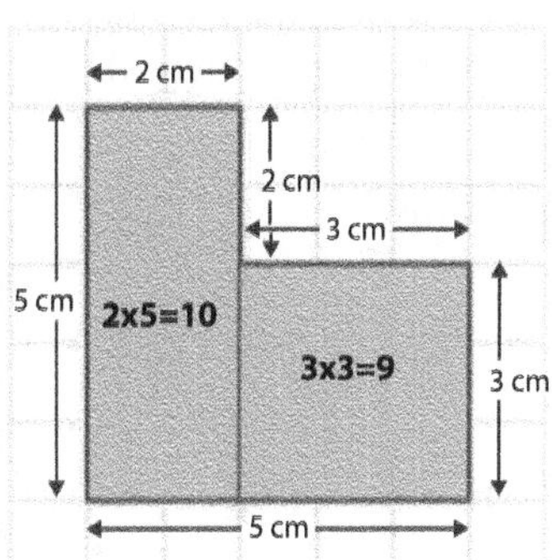

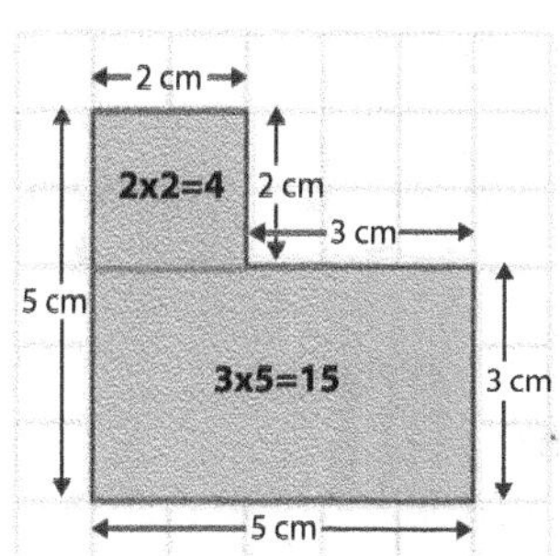

Area = 10cm^2 +9cm^2 = 19cm^2

Area = 4cm^2 +15cm^2 = 19cm^2

There are two general methods for finding the area of a composite shape.

1. *Additive Areas Method*

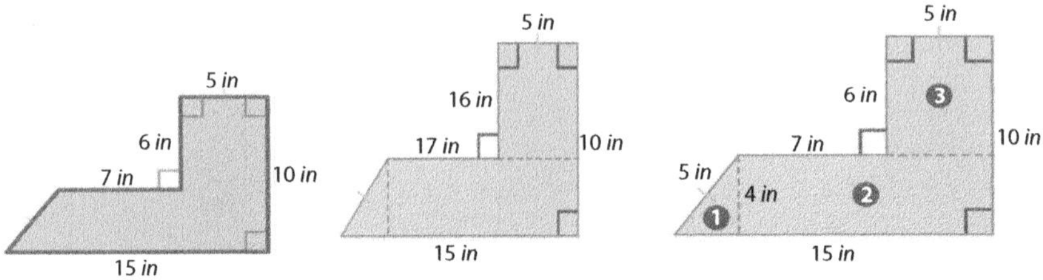

Here are a few steps to follow:

1. Divide the shape into one triangle and two rectangles (although it could have been broken up differently)
2. find the lengths of all the missing sides.
3. Find the separate area of each shape (one triangle and two rectangles).
4. Add the areas of all individual shapes to find the total area of the given shape.

Hint:
Shape (1) - right triangle

- Height = 10 - 6 = 4in (because the total height of the shape is 10 in, and the height above the triangle is given as 6 in.)
- Base = 15-(7+5) = 3in (because the total length of the base is 15 inches and other portions are 7 inches and 5 inches).

$$\text{Area (1)} = \frac{1}{2} \times 3 \times 4 = 6\ in^2$$

Shape (2) – rectangle

- Length =12 in
- Width = 4 in.

$$\text{Area (2)} = 12 \times 4 = 48\ in^2$$

Shape (3) – rectangle

- Length = 5 in
- Width = 6 in.

$$\text{Area (3)} = 5 \times 6 = 30\ in^2$$

Total area = Area (1) + Area (2) + Area (3)

$$6 + 48 + 30 = 84\ in^2$$

2. *Subtraction Areas Method*

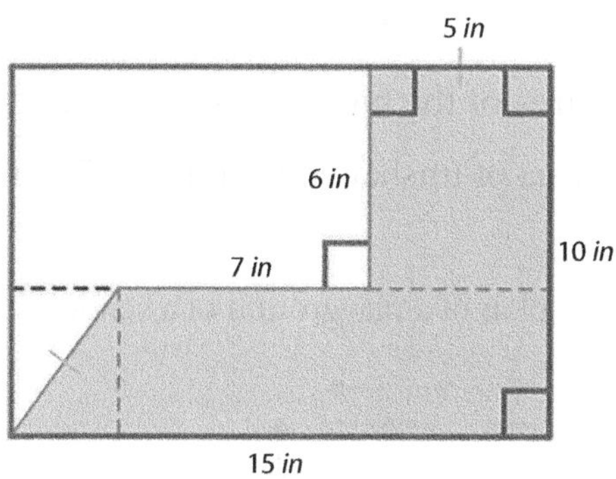

Observe the given shape.

Hint:
Shape (1) – Big rectangle (shades + unshaded) has the dimensions.

- Length =15 in
- Width = 10 in

$$\text{Area (1)} = 15 \times 10 = 150\ in^2$$

Shape (2) - The unshaded rectangle (white background) has the dimensions.

- Length =15 - 5 = 10 in
- Width = 6 in

$$\text{Area }(2) = 10 \times 6 = 60\ in^2$$

Shape (3) – The unshaded triangle (white background) has the dimensions.

- Height = 10 - 6 =4 in (The total height of the shape is 10 in., and the height above the rectangle is 6 in).
- Base = 3 in (by using the Pythagorean theorem).

$$\text{Area }(3) = \frac{1}{2} \times 4 \times 3 = 6\ in^2$$

Total area = area of the big rectangle – Area of the unshaded rectangle – Area of unshaded triangle $= 150 - 60 - 6 = 84\ in^2$

EXAMPLE 26: The plan of a playground of a school is shown below:

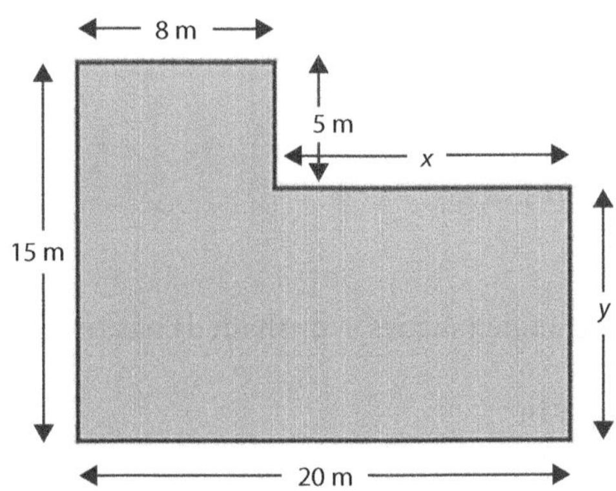

a. Find the length of x and y.
b. Find the Perimeter of the playground.

SOLUTION: In this case, we must calculate the 'missing lengths", 'x,' and 'y.'

a. From the figure

$$x = 20 - 8 = 12 \text{ m}$$

$$y = 15 - 5 = 10 \text{ m}$$

b. Perimeter of the play area

$$= 8 + 5 + 12 + 10 + 20 + 15 = 70 \text{ m}$$

EXAMPLE 27: In the given shape, Point O is the center of the semi-circle. Find all the missing sides for the shape and the Perimeter.

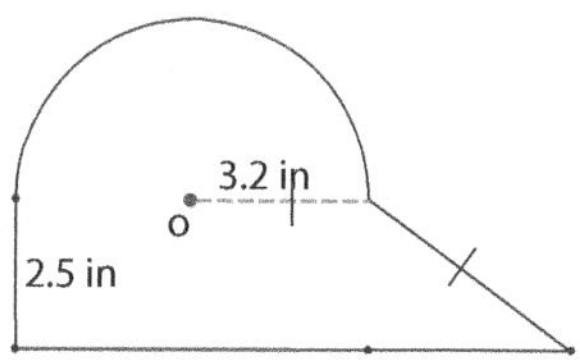

SOLUTION:

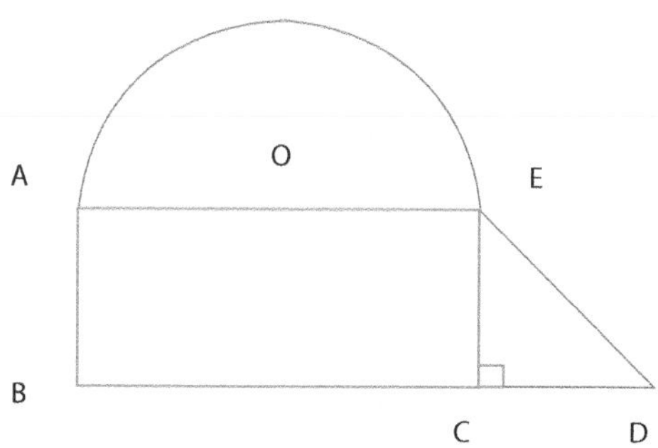

OE = r = 8.2 in
ED = DE = 8.2 in
AB = 2.5 in
AC = AE =2r =6.4
EC= AB = 2.5
In triangle ECD, Apply Pythagoras' theorem.

$$CD^2 = ED^2 - EC^2 = (3.2)^2 - (2.5)^2 = 10.24 - 6.25 = 3.99 = 4(\text{approx.})$$

CD =2 in
BD = 6.4 + 2 = 8.4 in
$AE = \pi r$ = 10.06 in
The perimeter of the shape is
AB + BD + ED + semicircle AE
2.5 + 8.5 + 3.2 + 10.06 = 24.16 in (approx.)

EXAMPLE 28: For the following figure, assume all the angles that look like right angles are right angles.

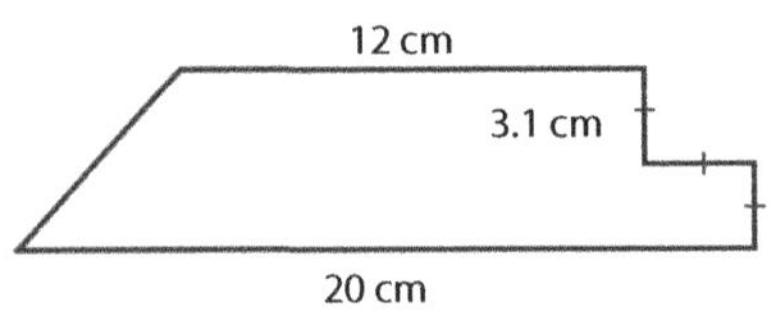

a. Find all the missing sides of the shape and the Perimeter.

SOLUTION:

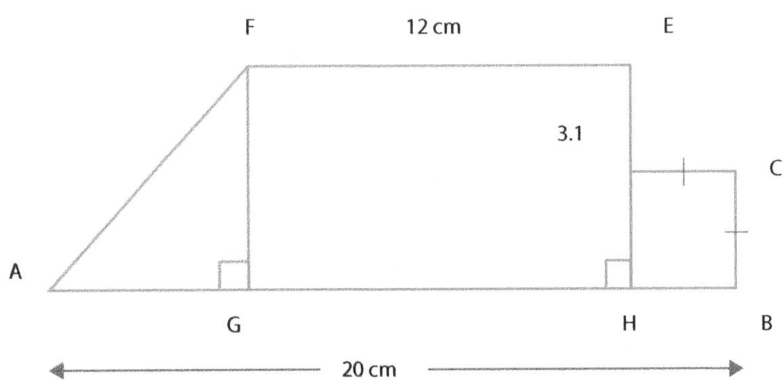

a. BC = CD = DE = 3.1 cm (HB = CD)

FG = EH = ED +DH = 3.1 +3.1 =6.2 cm
GH = FE =12 cm
AG= AB – GH -HB =20 -12 -3.1 = 4.9 cm
ΔAGF apply Pythagoras theorem

$$AF^2 = AG^2 + GF^2$$

$$AF^2 = (4.9)^2 + (6.2)^2$$

$$AF^2 = 24.01 + 38.44$$

$$AF^2 = 62.4$$

$$AF = \sqrt{62.4} = 7.9cm$$

b. Perimeter of the shape = AF + FD +ED + DG+ CB + BA

= 7.9+12+3.1+3.1++3.1+20 = 49.2 cm

EXAMPLE 29: Convert the following units as required (use the table above)

(a) 50 cm^2 in mm^2
(b) 4 hectares in m^2
(c) 12000 m^2 in hectares
(d) 2.5 hectares in acres

SOLUTION:

(a) 50 cm^2 in mm^2

50 cm^2 = 50 × 100 mm^2 = 5000 mm^2 (1 cm = 10 mm, 1 cm^2 = 100 mm^2)

(b) 4 hectares in m^2

4 hectares = 4 × 10000 m^2 = 40000 m^2 (1 hectare = 10000 m^2)

(c) 12000 m^2 in hectares

12000 m^2 = 1/10000 × 12000 hectare = 12/10 hectare = 6/5 hectare (10000 m^2 = 1 hectare)

(d) 2.5 hectares in acres

2.5 hectares = 2.5 × 100 acres = 250 acres(1 hectare = 100 acres)

EXAMPLE 30: Sameer painted the floor of the given shape at $0.10 per square meter. How much does Sameer earn by painting this area:

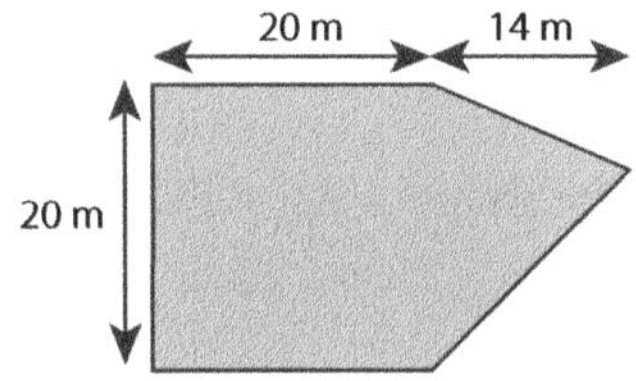

SOLUTION: First, divide the shape into two parts:

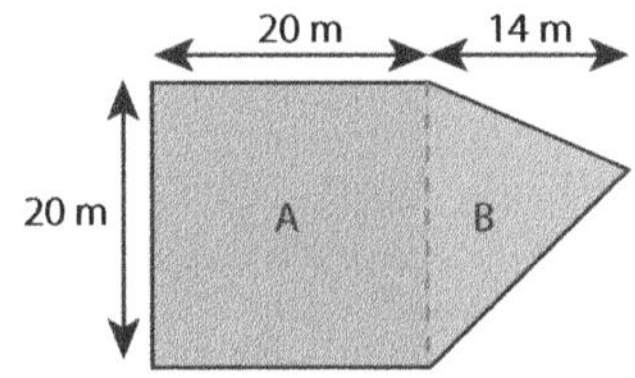

Shape A is a Square:
Area (A) = a^2

$$= 20 \times 20 = 400 \text{ m}^2$$

Shape (B) is a Triangle

$$\text{Area (B)} = \frac{1}{2} \times b \times h$$
$$= \frac{1}{2} \times 20 \times 14 = 140 \text{ m}^2$$

The Total Area = Area (A)+ Area (B)

$$= 400 + 140 = 540\text{m}^2$$

Sameer earns \$0.10 per square meter
Sameer earns = \$0.10 × 540m^2 = \$54

EXAMPLE 31: Find the Area of the shaded portion. (Hint: *Finding the Area of the whole circle and subtracting the area of the square is much simpler than trying to calculate the area of each of the four red pieces directly)*

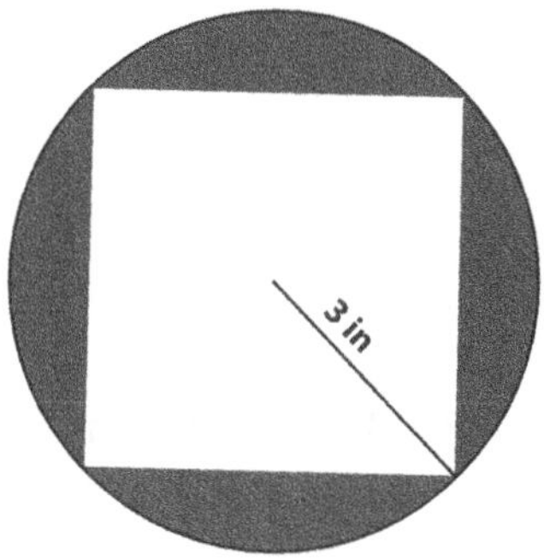

SOLUTION: Radius of circle(r)= 3 inch
Diagonal of square (d) =6 inch

Area of shaded portion = Area of Circle(full shape) - Area of Square(white background)

$$= \pi r^2 - \left(\frac{d}{\sqrt{2}}\right)^2$$

$$= \frac{22}{7} \times 3 \times 3 - \frac{6 \times 6}{2}$$

$$= 28.29 - 18 = 10.29 in^2$$

EXAMPLE 32: If the Area of a square is 144 in^2. Find the length of the side.

SOLUTION: Given- the area of the square = 144 in^2.

$$(\text{side})^2 = \text{Area of the square}$$

$$S^2 = 144$$

$$S = \sqrt{144} = 12 \text{ in}$$

EXAMPLE 33: If the Area of a square is 225 cm^2. Find its Perimeter?

SOLUTION: To determine the perimeter, we must first find the length of the side of the square. Then use that side to determine the perimeter.

$$(\text{side})^2 = \text{Area of the square}$$

$$S^2 = 225$$

$$S = \sqrt{225} = 15$$

Now, Perimeter = 4 × side

$$P = 4 \times 15 = 60 \text{ cm}$$

EXAMPLE 34: The total Area of a park is 300 m^2. If the park length is times its width, what is the difference between the length and width of the park?

SOLUTION: Let the width of the park be x

$$\text{Length of the park } \frac{4x}{3}$$

$$\text{Area of park} = l \times b$$

$$\frac{300 \times 3}{4} = x^2$$

$$x = \sqrt{\frac{300 \times 3}{4}} = \sqrt{\frac{900}{4}} = \frac{30}{2} = 15 \text{ m}$$

Width of the park =15 cm, Length of the park = 20 cm
Difference between the dimensions = 5 m

EXAMPLE 35: Calculate the height (h) of the rhombus

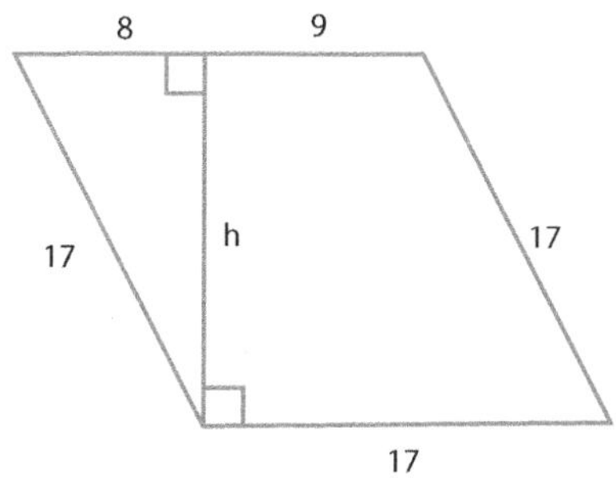

SOLUTION:

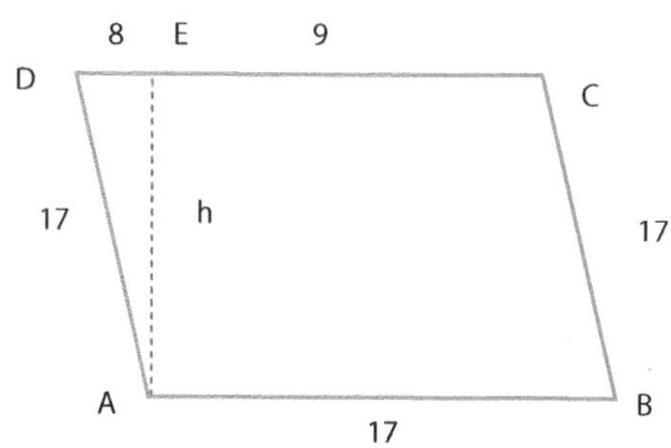

In AED applying the Pythagorean theorem

$$AD^2 = AE^2 + DE^2$$
$$17^2 = h^2 + 8^2$$
$$289 = h^2 + 64$$
$$h^2 = 289 - 64$$
$$h^2 = 225$$
$$h = 15$$

Height=15m

EXAMPLE 36: Calculate the Area of the given shape:

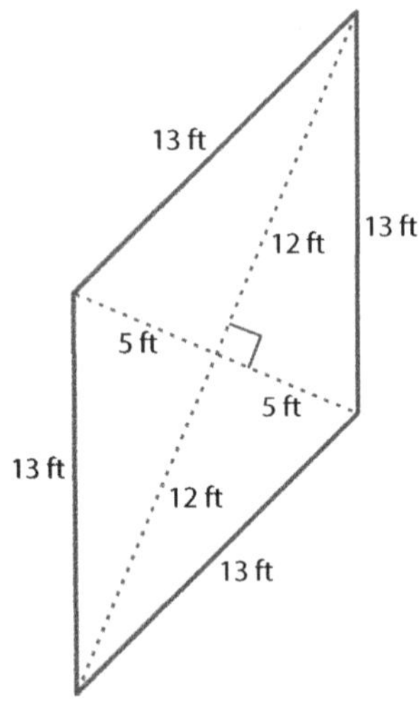

SOLUTION: $d_1 = 10$ ft $\quad d_2 = 24$ ft

$$\text{Area of Rhombus} = \frac{1}{2} \times d_1 \times d_2$$

$$\text{Area of Rhombus} = \frac{1}{2} \times 10 \times 24 = 120\ ft^2$$

EXAMPLE 37: The area of the given parallelogram is 132 inch^2, and its base is 11 inches; find its height.

SOLUTION:

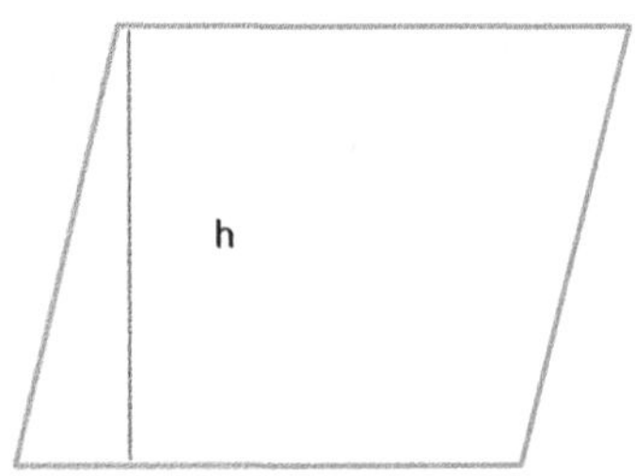

As given, Base = 11 in, Area = 132 sq in
Let the height of the parallelogram = h

We know that Area = Base × Height
Therefore, 132 = 11 × h

$$A = \frac{132}{11}$$

h=12 in
Height of parallelogram=12 in

EXAMPLE 38: A rhombus-shaped floor of the house is made up of 4000 tiles. If each of its diagonals is 25 cm and 20 cm, Find the expenditure of polishing the floor at the rate per m^2 is $5.

SOLUTION: Given that the diagonals are = 20 cm and 25 cm

$$\text{Hench the area of each tile} = \frac{1}{2} \times 25 \times 20 = 250 \text{ cm}^2$$

Area of 4000 tiles = 4000 × 250 cm^2
= 1000000 cm^2
= 1000000/10000 cm^2
= 100 m^2

Expenditure for 1 m^2 = $5
Total expenditure = $5 × 100 = $500.

EXAMPLE 39: The base of an isosceles triangle is 24 cm. If each of its equal sides is 13 cm. Find its area.

SOLUTION: The three sides of triangle are a=13 cm, b=13cm, c= 24cm

$$s = \frac{a+b+c}{2}$$

$$S = \frac{13+13+24}{2} = \frac{50}{2} = 25$$

$$\text{Area of triangle} = \sqrt{25(25-13)25-13)(25-24)} = \sqrt{25 \times 12 \times 12 \times 1}$$

$$\text{Area of triangle} = \sqrt{5 \times 5 \times 12 \times 12} = 5 \times 12 = 60cm^2$$

EXAMPLE 40: If the perimeter of a triangle is 54 cm, and two sides measure 12 cm, and 25 cm, Find the area.

SOLUTION: The perimeter of a triangle is = 54 cm

$$2s = 54, s = 27$$

$$a = 12, b = 25$$

$$c = p - (a + b)$$

$$54 - 37 = 17 \text{ cm}$$

$$\text{Area of triangle} = \sqrt{s(s-a)(s-b)(s-c)}$$

$$\text{Area of triangle} = \sqrt{27(27-12)(27-25)(27-17)}$$

$$= \sqrt{27(15)(2)(10)}$$

$$= \sqrt{3\times3\times3\times3\times5\times2\times5\times2} = 3\times3\times5\times2 = 90 \text{ cm}$$

EXAMPLE 41: If the perimeter of a triangle is 144cm, and the lengths of its sides are in the ratio 3: 4: 5.
Find (i) The area (ii) The height corresponding to the longest side.

SOLUTION: Let the three sides of a triangle are $3x$, $4x$, $5x$
Given the perimeter of the triangle =144 cm

$$3x + 4x + 5x = 144$$

$$12x = 144$$

$$x = 12$$

Therefore, the sides of the triangle are

$$a = 3\times 12 = 36 \text{ cm},\ b = 4\times 12 = 48 \text{ cm},\ c = 5\times 12 = 60 \text{ cm}$$

$$\text{Area of triangle} = \sqrt{s(s-a)(s-b)(s-c)}$$

$$\text{Area of triangle} = \sqrt{72(72-36)(72-48)(72-60)}$$

$$\text{Area of triangle} = \sqrt{72\times 36\times 24\times 12} = \sqrt{72\times 36\times 2\times 12\times 12}$$

$$= \sqrt{72\times 72\times 12\times 12} = 72\times 12 = 864cm^2$$

The longest side is = 60cm.
consider the height corresponding to the longest side is hcm

The area of the triangle $= \frac{1}{2} base \times height$

$$864 = \frac{1}{2}\times 60\times h$$

$$h = \frac{864\times 2}{60} = 28.8 \text{ cm}$$

EXAMPLE 42: Calculate the area of the shaded regions in the given figure.

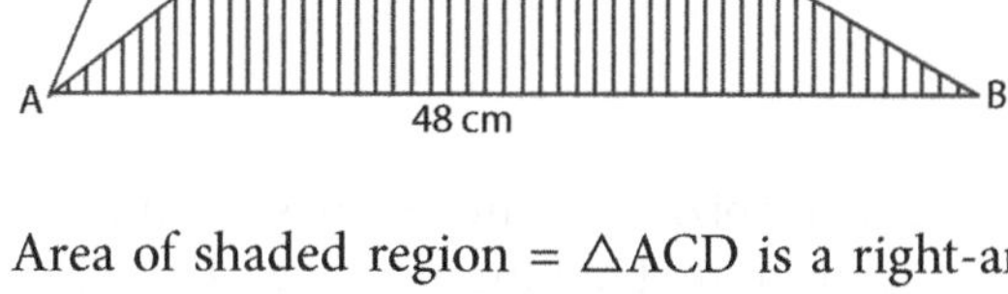

SOLUTION: Area of shaded region = ΔACD is a right-angled triangle. With CD=12, AD=16

By Pythagoras theorem

$$AC^2 = CD^2 + AD^2$$
$$AC^2 = 12^2 + 16^2$$
$$AC^2 = 144 + 256$$
$$AC^2 = 400$$
$$AC = 20$$

$$\text{Area of triangle ACD} = \frac{1}{2} base \times height$$
$$\text{Area of triangle ACD} = \frac{1}{2} \times 12 \times 16$$
$$\text{Area of triangle ACD} = 96cm^2$$

For △ABC.

$$\text{a} = 52\ cm,\ \text{b} = 20\ cm,\ \text{c} = 48\ cm$$
$$\text{s} = \frac{a+b+c}{2} = \frac{52+20+48}{2} = 60$$
$$\text{Area of triangle} = \sqrt{s(s-a)(s-b)(s-c)}$$
$$\text{Area of triangle} = \sqrt{60(60-52)(60-20)(60-48)} = \sqrt{60 \times 8 \times 40 \times 12}$$
$$\text{Area of triangle} = \sqrt{10 \times 10 \times 48 \times 48} = 480\ cm^2$$
$$\text{Area of shaded region} = 480 - 96 = 384\ cm^2$$

EXAMPLE 43: A circular pizza of radius 8 cm is cut into six pieces of equal size, as shown in the diagram. Find the area of each piece.

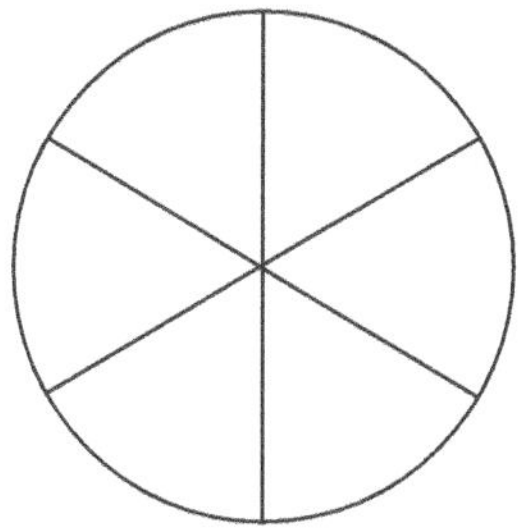

SOLUTION: The area of the circle is calculated as $A = \pi r^2$

$$A = 3.14 \times 8^2 = 3.14 \times 8 \times 8 = 200.96 cm^2$$

Thus, the area of each part of the circle is $\frac{A}{6} = \frac{200.96}{6} = 33.49 cm^2$

EXAMPLE 44: A circle has a circumference of 440 m. What will be its area?

SOLUTION: To find the area of the circle, we need to know its radius.
By using the circumference formula,

$$C = 2\pi r = 440$$

$$= 2 \times \frac{22}{7} \times r = 440$$

$$r = \frac{440 \times 7}{2 \times 22} = 70\ m$$

Now, we can find the area of a circle as $A = \pi r^2 = \frac{22}{7} \times 70 \times 70 = 15400\ \text{m}^2$

EXAMPLE 45: A triangular region having an area of 224 m² and a base of 28 m. Find its height.

SOLUTION: Given, base = 28 m and area = 224 m²

$$\text{Area of the triangle} = \frac{1}{2} \times b \times h$$

$$H = (2 \times \text{Area}) \div b$$

$$\therefore H = (2 \times 224) \div 28 = 16\ \text{m}$$

Hence the height of the triangular region = is 16m.

EXAMPLE 46: A goat is tethered with a rope of length 28 m at the corner of the rectangular field of dimensions 70 m ×52 m. finds the area of the field where the cow can graze.

SOLUTION:

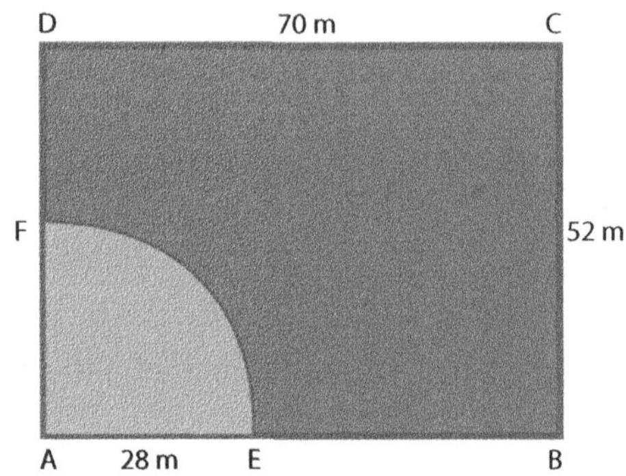

From the figure, we can understand the problem.

Since the goat is tethered with a rope at the corner of the field so she can only graze in a quarter of a circle, as shown in the figure.

So, the total area she can graze is the quadrant of a circle.

$$A = \frac{\pi r^2}{4} = \frac{22\times28\times28}{7\times4} = 22\times28 = 616\, m^2$$

A goat can graze in $616\, m^2$ area.

EXAMPLE 47: Find the Area of the quadrant if its radius is 3.5 cm.

SOLUTION: Here r = 3.5 cm and $\pi = \frac{22}{7}$

$$= \frac{1}{4}\times(\frac{22}{7})\times(3.5)^2 = \frac{1}{4}\times(\frac{22}{7})\times3.5\times3.5$$

$$= \frac{1}{4}\times22\times0.5\times3.5 = \frac{1}{2}\times11\times0.5\times3.5$$

$$= 5.5\times0.5\times3.5 = 18.375\ \text{cm}^2$$

Level 1: Exercise 3.1 Practice Questions

1. If the perimeter of a square is 44 cm. Find its area.
2. Aisha went to a playground 25 m long and 15 m wide. She took one complete round of it. Find the total distance covered by her.
3. Kevin wants to fence his rectangular park, whose dimensions are 12 m × 6 m. Find the cost of fencing at the rate of $3 per meter.
4. Rectangular land measures 0.55 km by 0.35 km. Each side of the land will be fenced with three rounds of wire. Find the length of the wire needed.
5. Calculate the perimeter of the triangle whose sides are:
   a. 10 m, 18 m, and 22 m
   b. 20 cm, 20 cm, and 15 cm
6. The perimeter of an isosceles triangle is 70 cm. Find the third side if the two equal sides are 21 cm each.
7. Calculate the perimeter of the following shapes:
   a. A triangle whose sides are 30 cm, 40 cm, and 50 cm.
   b. An equilateral triangle of a side length of 7 cm.
   c. An isosceles triangle whose equal sides are 9 cm each and a third side of 7 cm.
8. The sides of a triangle measure 13cm, 14cm, and 15cm. Find its area.
9. Find the perimeter of the following figures

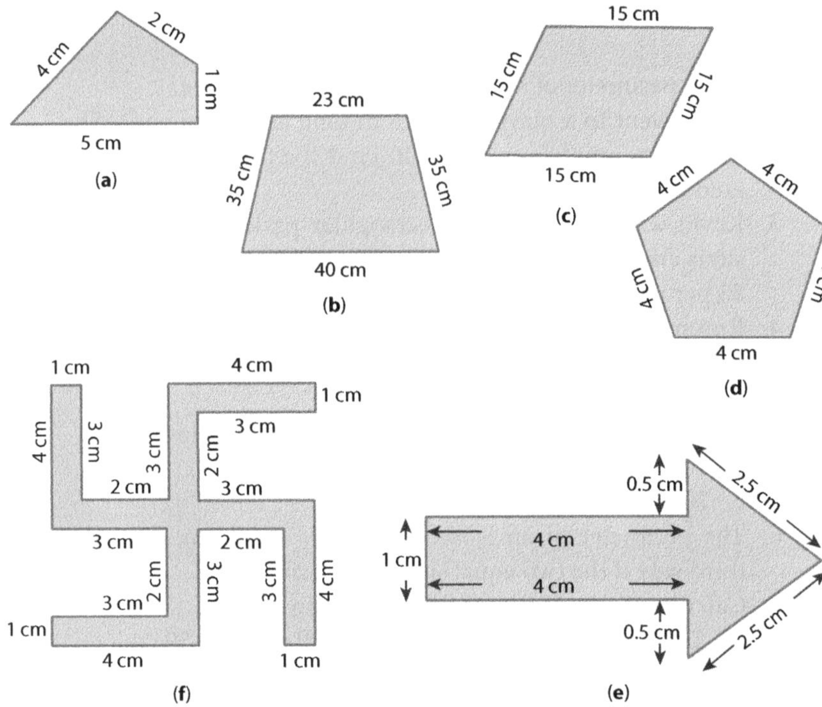

10. Find the perimeter of the given shape.

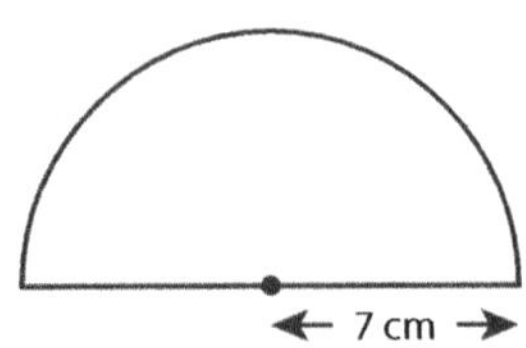

11. Calculate the circle's circumference with a diameter of 14 cm.
12. Sarah's Garden is shaped like a quarter-circle with a radius of 3.5 meters. Calculate the total length along which he needs to lay the logs.
13. If the area of two squares is in the ratio of 81:25, what is the ratio of their perimeters?
14. In a square field, the diagonal is 24 m. Find the area of the square field.

15. Think and answer: What happens to the area of the rectangle if:
    a. Length is doubled, and breadth remains the same.
    b. Length is doubled, and breadth is split in half.
    c. Breadth is tripled, and length is doubled.
16. Calculate the base of a triangle with an altitude of 17 cm and an area of 0.7 m^2.
17. Calculate the area of an isosceles right-angled triangle of equal sides 12 cm each.
18. Find the area of the following parallelogram.

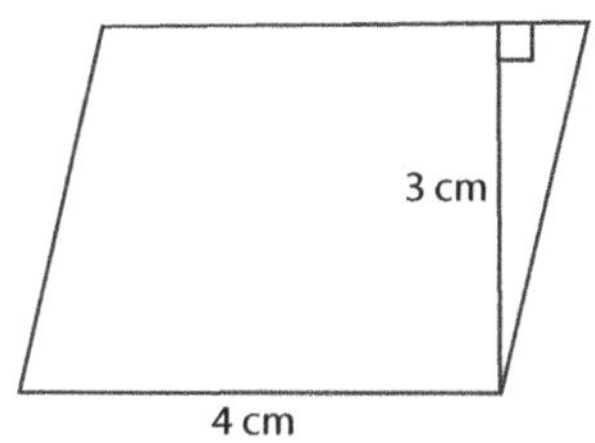

19. Find the area of a rhombus in which each side = 10 cm and height = 7.5 cm.
20. The parallel sides of a trapezium are 26 cm and 30 cm, and its height is 8 cm. What is the area?
21. Find the area of the given figure.

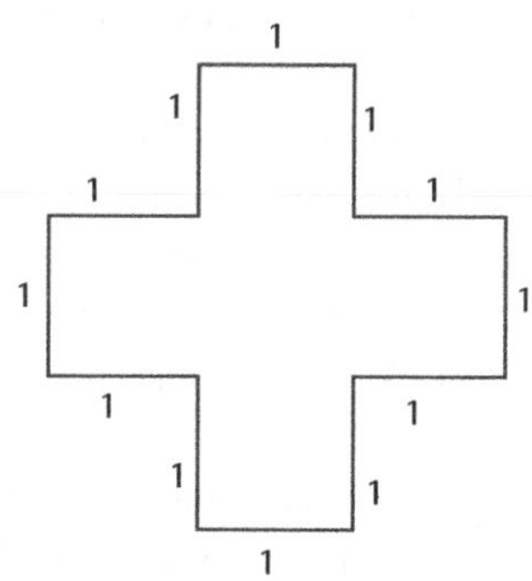

22. Find the area of the given figure.

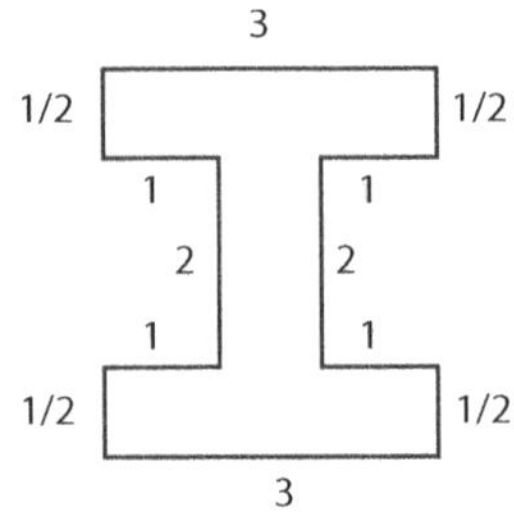

23. Calculate the area of a circle whose circumference is 132 m.
24. What is the area of a semi-circle if its perimeter is 144 cm?

Level 2: Exercise 3.2 Practice Questions

1. The dimensions of rectangular land are 1.5 km ×2 km. The whole land is to be fenced twice. Finds the length of the wire needed.
2. Ayesha walked 10 times around a square field of side 75m. Her friend Sarah walked 8 times around another rectangular field with a length of 150m and a breadth of 50m. Calculate who covers more distance. By how much?
3. If the length of a rectangular park is four times its breadth. If the park's perimeter is 900m, Find the park's length.
4. If the length of the rectangular construction field is 11 ft more than its width and the perimeter of the field is 162 ft, what will be its length and width?
5. If the length and width of a rectangular car parking area of a mall are in the ratio 7:5. If its perimeter is 580 m, calculate its dimensions.
6. As shown in the figure, three squares are attached at the mid-point of the side of the square to its right. Find the perimeter of the complete figure.

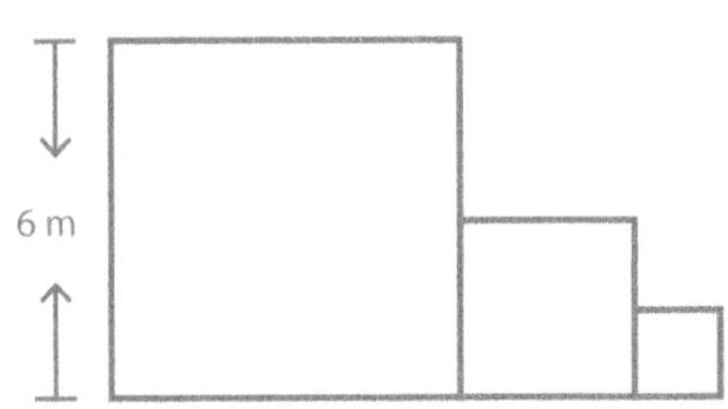

7. Calculate the area of the quadrilateral ABCD shown in the figure below using Heron's formula.

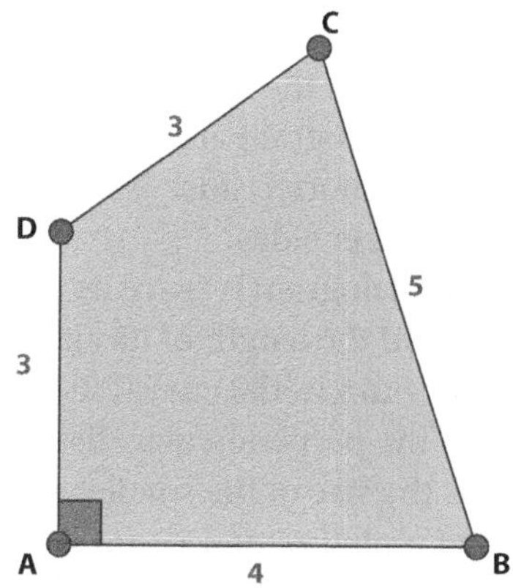

8. The perimeter of a rhombus-shaped land is 200 m, and one of its diagonals is 60 m. Find the area of the land.
9. If parallel sides of a trapezium-shaped parking field are 20 m and 12 m, and its height is 10 m. Find the area of the parking field.
10. Find the area of the following shapes:

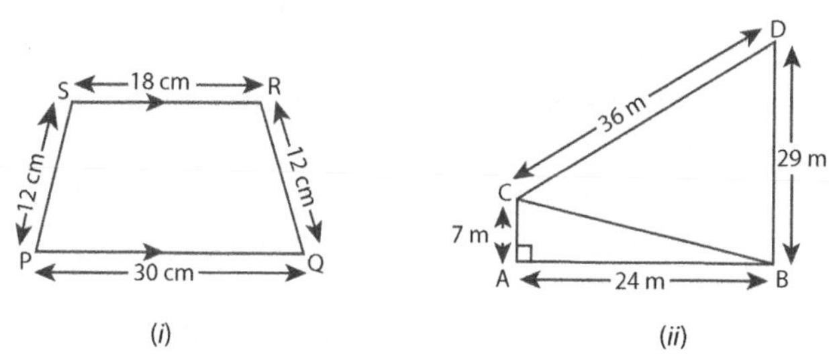

11. Calculate the side of a regular pentagon whose perimeter is 95 cm.
12. Calculate the number of revolutions of a wheel whose radius is 21 cm, which can rotate to 924 m.
13. If the length of the minute hand of a circular clock is 10 cm. Calculate how far the tip of the minute hand moves in 1 hour.
14. If the length of a rectangular swimming pool is 22 ft and its width is half its length, find the width and area of the pool.

15. The floor is 6 m long and 4 m wide. A square carpet of side 1.5 m is placed on the floor. Find the area of the floor without a carpet.
16. If the base and height of a triangle are in the ratio 5: 2, and its area is 250 m^2. Calculate the height and base of the triangle.
17. If the sides of a parallelogram are 14 cm and 10 cm, and the distance between its shorter sides is 8 cm. What is the distance between its longer sides?
18. If the base of a parallelogram is twice its altitude, and its area is 242 cm2, then find the length of its altitude.
19. The area of a trapezium is 450 cm^2. If its parallel sides are in the ratio 2: 3 and the perpendicular distance between them is 10 cm. What is the size of the smaller sides?
20. Adam is painting a picture of a rising sun, which is exactly half of a circle. The base of the half-circle is 28 inches long. What quantity of paint does he need to buy if each 100 ml of paint can cover 45 inches of the space? (Write your answer to the nearest whole) (take $\pi = 3.14$)

Level 3: Exercise 3.3 Practice Questions

1. 24 squares are joined to form a rectangle with the smallest perimeter. Find the perimeter of the rectangle.
2. The perimeter of the two squares is 4:3, respectively. The sum of the perimeters of the squares is 56 m. Find the perimeter of both squares.
3. The base of a rectangular field is half of its height. If the cost of fertilizing the field at $95 per hectare is $2375, find its base and height.
4. Find the area of the following compounded shapes. (The measures are given in centimeters)

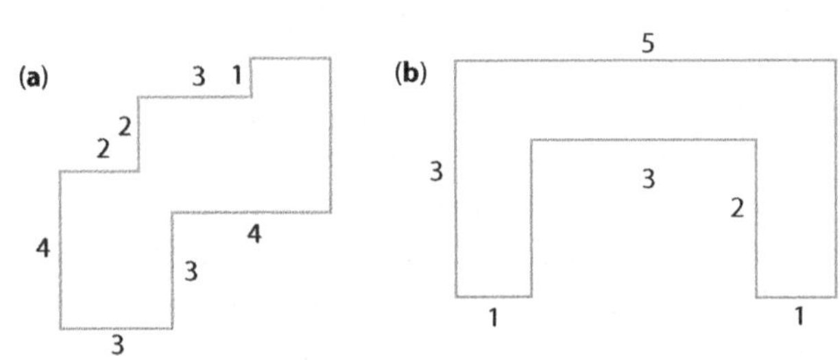

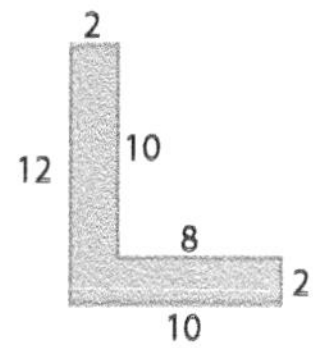

(a)

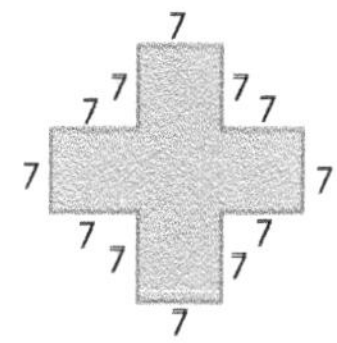

(b)

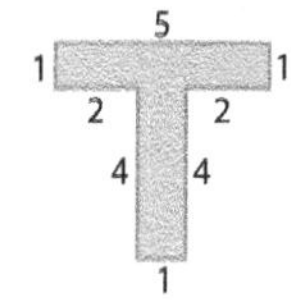

(c)

5. Calculate the area of the shaded region of the following compounded shapes.

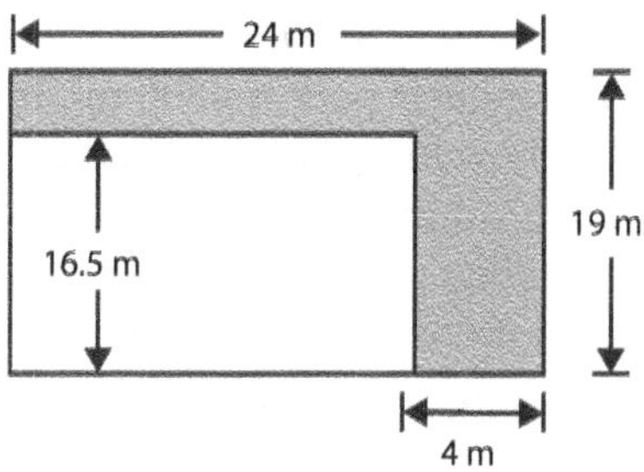

6. Work out the area of each of these shapes.

(a)

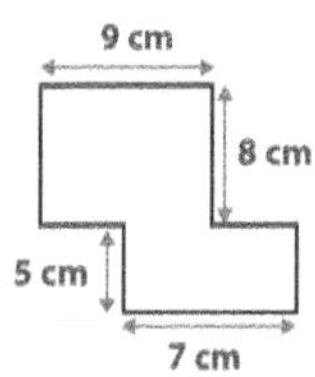

(b)

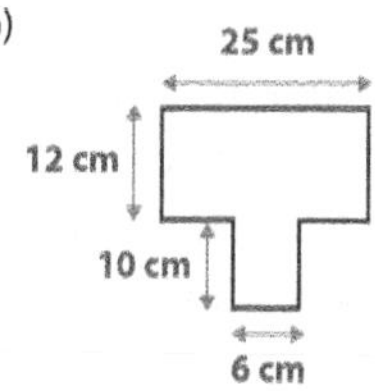

(c)

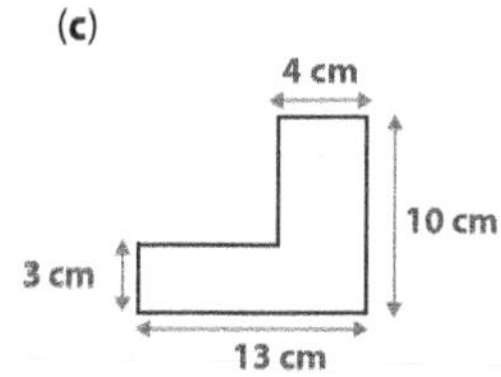

(d)

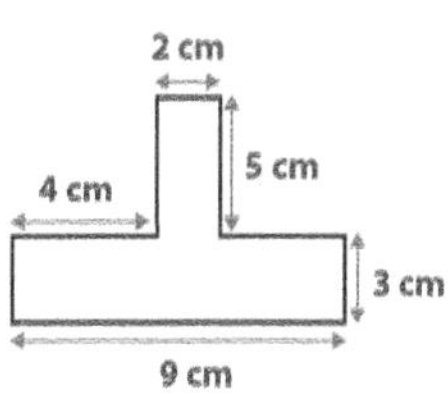

(e)

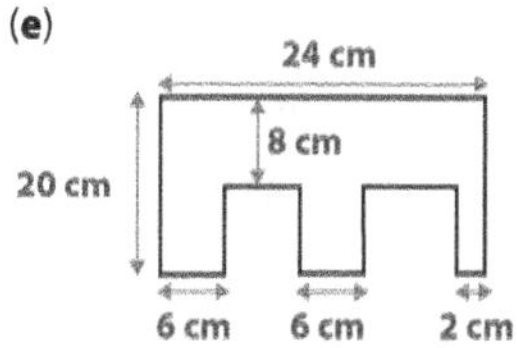

(f)

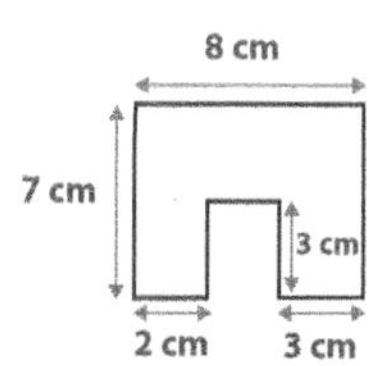

7. Find the area of the shaded region.

(a) (b) (c)

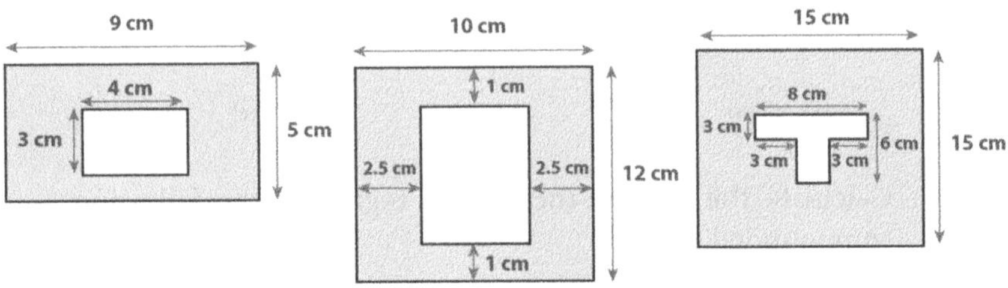

8. The area of a rectangular park is 4800 m^2. If one of its sides is 80m, calculate how long Aisha will take to cross the field diagonally at 25 m/minute.
9. A 20m long walking pathway has an area of 240 m^2. If municipal corporations want to widen the walking path, they must increase their area by $1\frac{2}{3}$ times the original area by increasing only its breadth. Calculate the new perimeter of the walking track.
10. A rectangular field has dimensions of 90 m by 70 m. Inside the field, a cycling track of a uniform width of 2.5 m is to be built all around. All other areas of the field are to be laid with grass. Find the total expenses of constructing the cycling track at \$8 per m^2 and laying the grass at \$2 per m^2.
11. Refer to the below figures. Find the total area.

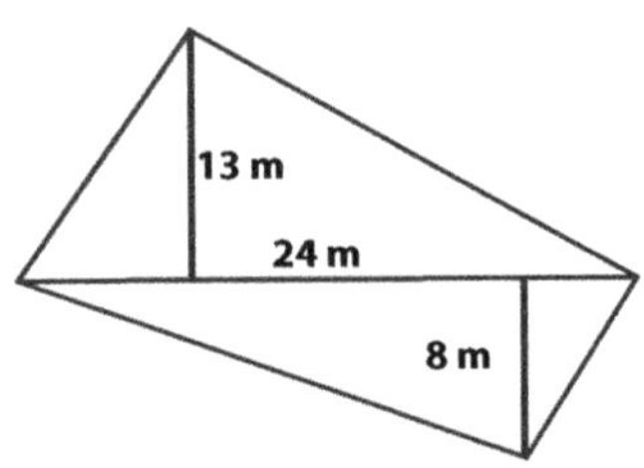

12. In the figure below, calculate the area of the shaded portion.

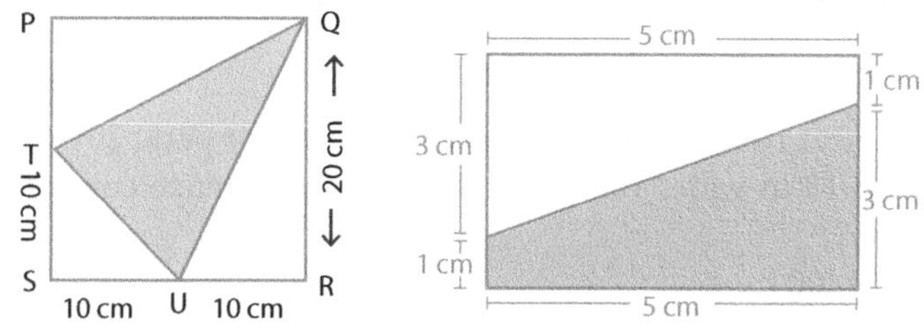

13. The area of the trapezium is 385 cm^2, and its height is 10 cm. Find the length of its parallel of a trapezium if they are in the ratio of 7:4.
14. There are two new offers at Pizza Hut. First, any square pizza side of 45 cm costs \$15, and any circular pizza with a diameter of 50 cm costs \$16. Which pizza do you want to buy as the best deal?

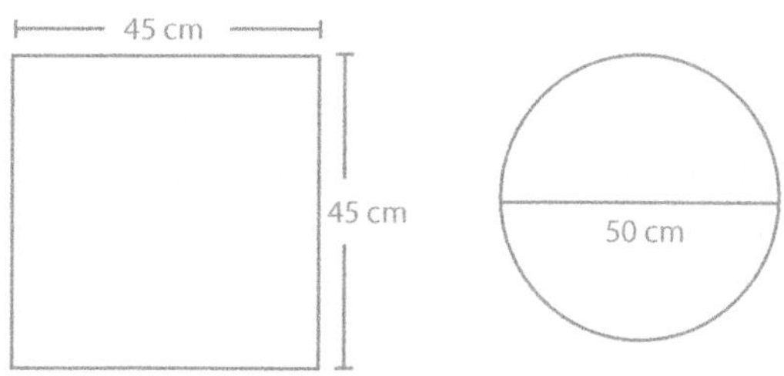

15. Aisha runs on a track that is semi-circular at the ends and rectangular in the middle. If she runs one full round around the track, find the distance covered by her and the total area of the track (Refer to diagram).

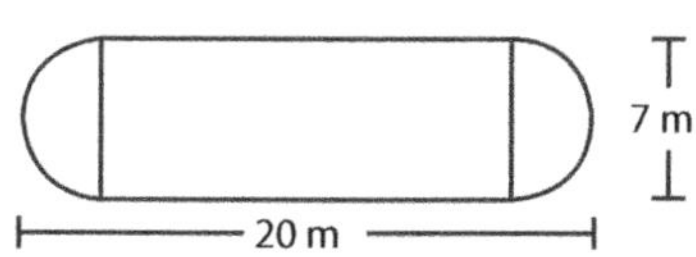

3.6 Check Your Knowledge (MCQ)

1. Sameer runs 5 rounds around a rectangular park, 20 m long and 15 m wide. What will be the total distance covered by him?
   a. 1400 m
   b. 700 m
   c. 350 m
   d. 2800 m

2. Two squares of size 1 cm^2, and 3 cm^2 are joined together. What is the perimeter of the new shape?
   a. 13cm
   b. 14 cm
   c. 15cm
   d. 16 cm

3. A rectangular cable of length 25 cm and breadth 16 cm is bent as a square. What is the side of the square?
   a. 10 cm
   b. 20 cm
   c. 30 cm
   d. 41 m

4. If the diagonal of a square is 32 cm. What is its perimeter? (Take $\sqrt{2}=1.41$)
   a. 93.42 cm
   b. 90.24 cm
   c. 92.24 cm
   d. 87.65 cm

5. The perimeter of a triangle is 12 cm if its two sides are 5 cm and 4 cm. The third side has the length will be?
   a. 1 cm
   b. 2 cm
   c. 3 cm
   d. 6 cm

6. The perimeter of the figure below is:

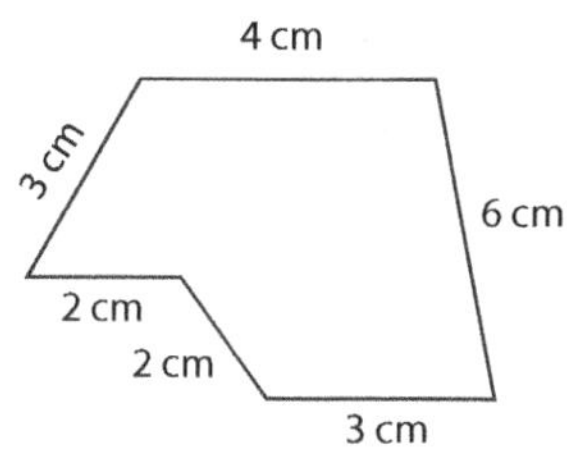

   a. 15 cm
   b. 10 cm
   c. 24 cm
   d. 20 cm

7. The perimeter of a regular hexagon of side 1 m is
   a. 4 m
   b. 2 m
   c. 3 m
   d. 6 m

8. The circumference of a circle is 37 m more than its radius. What is the circumference of the circle?
   a. 44 m
   b. 7 m
   c. 154 m
   d. 88 m

9. The radius of the circle is 28 cm which is formed by a wire. If the same wire is folded and forms a square, what will be the side of the square?
   a. 66 cm
   b. 44 cm
   c. 22 cm
   d. 11 cm

10. There are two square-shaped fields of perimeters 24 meters and 32 meters. What is the perimeter of a third squared field whose area is equal to the sum of the areas of the two squared fields?
    a. 50 m
    b. 40 m
    c. 30 m
    d. 60 m

3.7 Activity/Experiment

Squares and rectangles

On this board, there are 16 dots. Find out!

1. How many total squares can you make? If the distance between any two dots (vertical or horizontal) is 2cm, find out the area and perimeter of each square.
2. How many triangles can you make? If the distance between any two dots (vertical or horizontal) is 2cm, find out the area and perimeter of each rectangle.

Use the below table for your work.

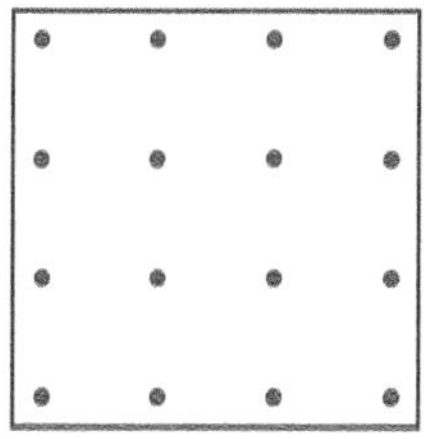

Total number of squares	
Total number of rectangles	

Squares/rectangles	**Area**	**Perimeter**
Square type 1		
Square type 2		
Square type 3		
Rectangle type 1		
Rectangle type 2		
Rectangle type 3		

4

Three-Dimensional Geometry

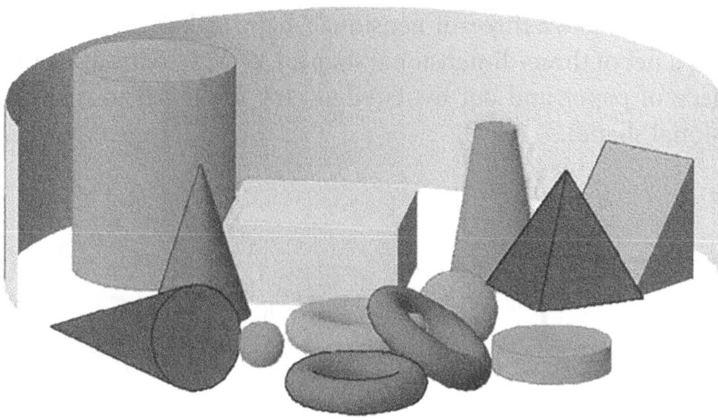

Objectives: This chapter

- *Describes three-dimensional (3D) geometric figures.*
- *Explains the nets of three-dimensional figures.*
- *Defines the surface area and volume of three-dimensional shapes.*
- *Solve different levels of problems related to volume and surface areas of 3D shapes.*

4.1 Introduction

Geometry is the branch of mathematics that measures two-dimensional and three-dimensional shapes or figures. The previous topic of Chapter 3 was two- dimensional shapes; here, we will learn rules and methods to measure the volume and surface of any three-directional shape.

The only difference between 2-dimensional and three-dimensional shapes is the third dimension which is Depth/height along with length

Ritu Shrivastava, Ramakant Bhardwaj and Satyendra Narayan. Mathematics for Engineers, (121–170)

and width. Any two-dimensional shape has length and width, whereas a three-dimensional solid shape also has depth, length, and breadth.

4.2 Nets of Three-Dimensional Shapes

Observe the below figure. If you look at it carefully, you can identify that this shape can make a three-dimensional shape called a cylinder. It is also known as a net of three-dimensional shape. (As an activity, draw this shape on a piece of paper and cut by its edges, try folding it to make a three-dimensional shape).

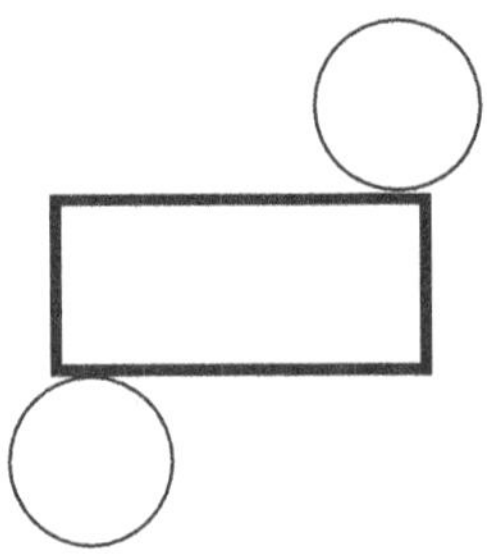

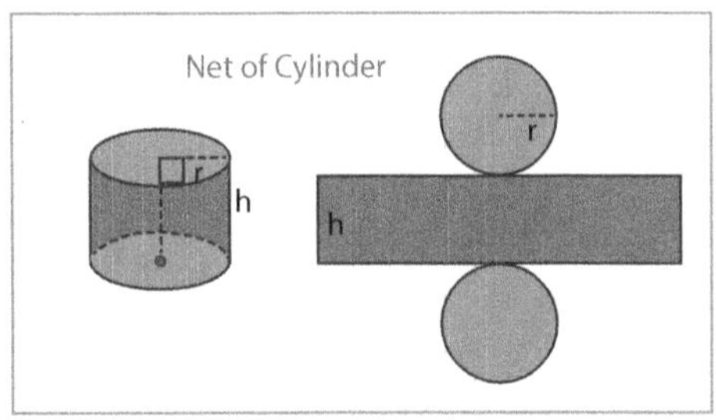

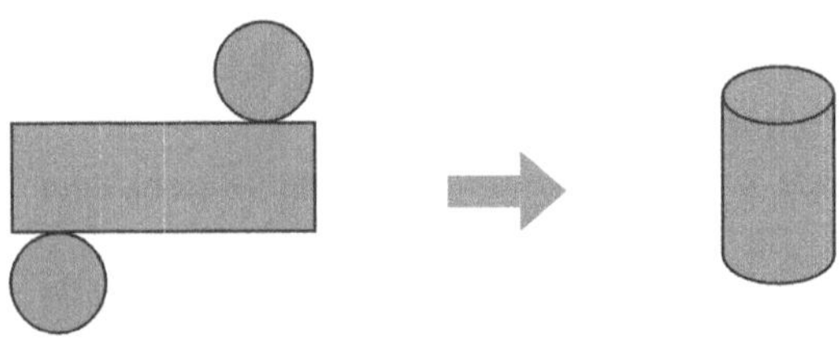

This net of the cylinder has two circles (top and bottom and a rectangle as its curved surface)

EXAMPLE 1: Identify the correct numbers for the empty face of the dice. (Dice are cube in shape and have 6 faces)

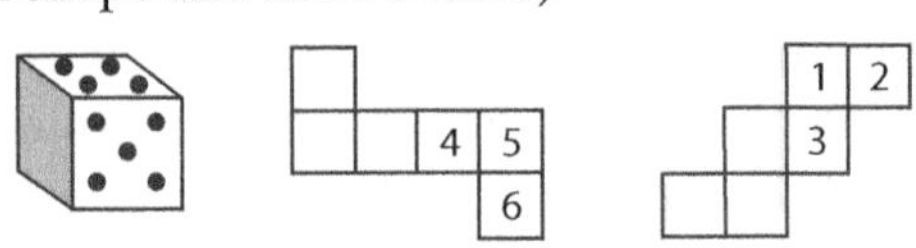

SOLUTION:

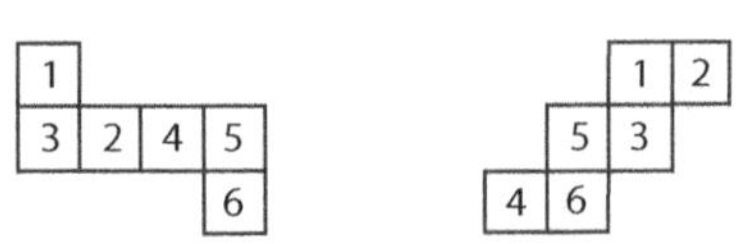

EXAMPLE 2: Match the correct nets.

(a) 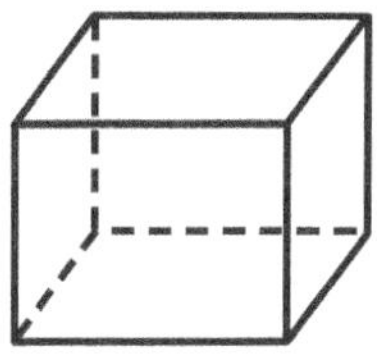 (i) 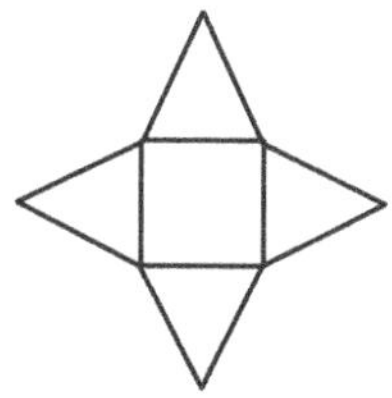

(b) 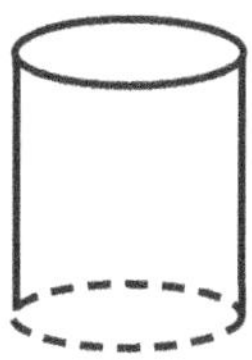 (ii) 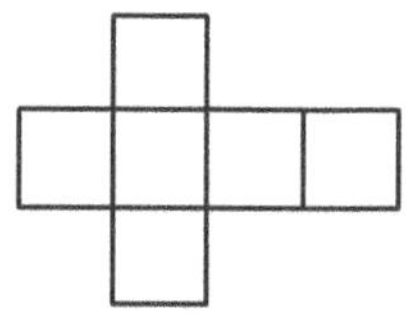

(c) 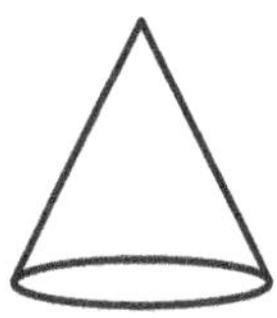 (iii) 

(d) 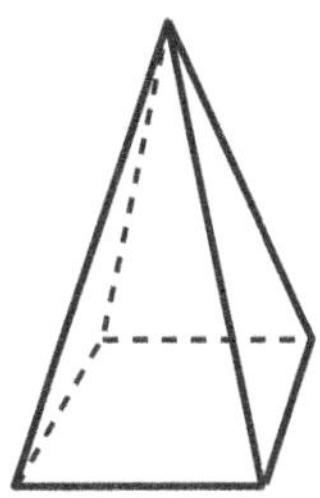 (iv) 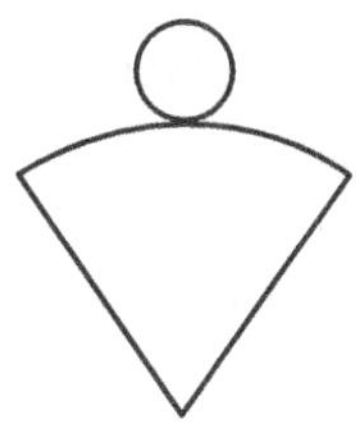

SOLUTION: understand this example by the below colored figures.

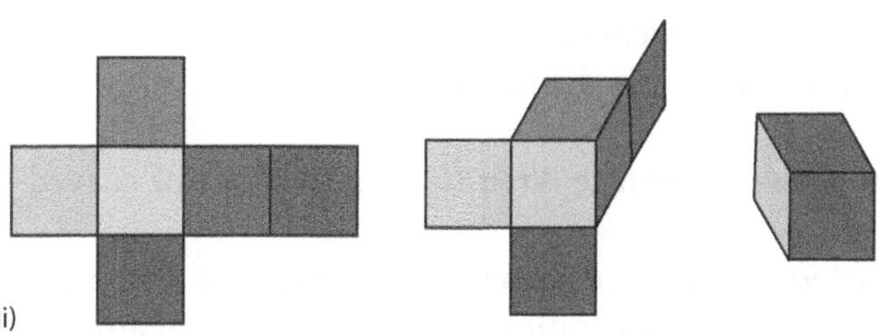

(**a**) (ii)

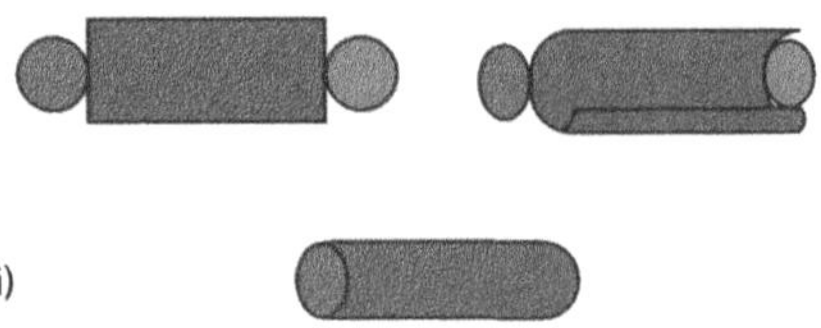

(b) (iii)

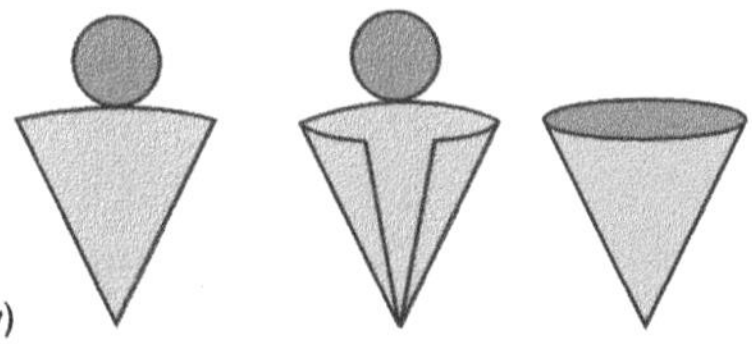

(c) (iv)

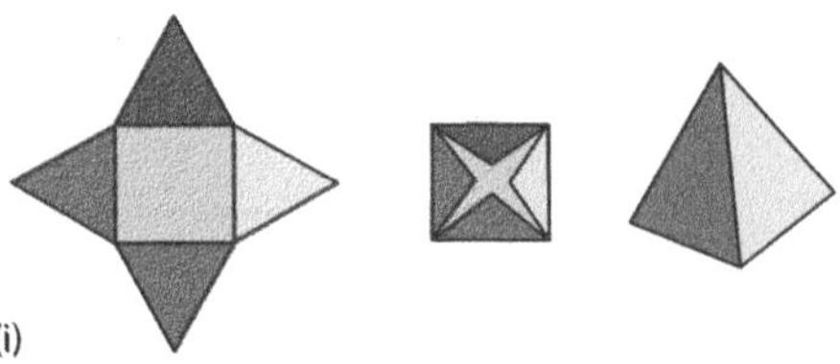

(d) (i)

4.3 Surface Area and Volume of Three-Dimensional Shapes

Surface Area and volume are the two very essential measures for defining all 3D shapes.

Surface area: The total area of the surface of the three-dimensional object is known as its Surface area. Surface area is the 2D measure of any 3D shape. Hence, the surface area is measured in square units

Nets are the best way to calculate the surface area. There are three different types of surface areas of any solid surface:

Let's learn a few more definitions,

The Curved Surface Area (CSA): The sum of curved surface areas.

The Lateral Surface Area (LSA): The sum of all the curved and flat surfaces, excluding base and top areas(if any).

The Total Surface Area (TSA :) is the sum of the areas of all the surfaces, including the base and top of the three-dimensional object.

Volume:

The total space occupied by any three-dimensional shape or solid object is known as its volume. Volume is measured in cubic units. The capacity of an object can be considered as the volume of that object. It refers to the amount of space an object occupies, and capacity refers to a container's quantity.

Note: If a container holds 1000 cm^3 of water, then the capacity of the container is 1000 cm^3, and the unit to measure capacity is in liters.

1 ml = 1 cm^3

1 L = 1000 cm^3

Thus, 1 m^3 = 1000000 cm^3 = 1000 L

Any three-dimensional figure contains **faces, edges, and vertices.**

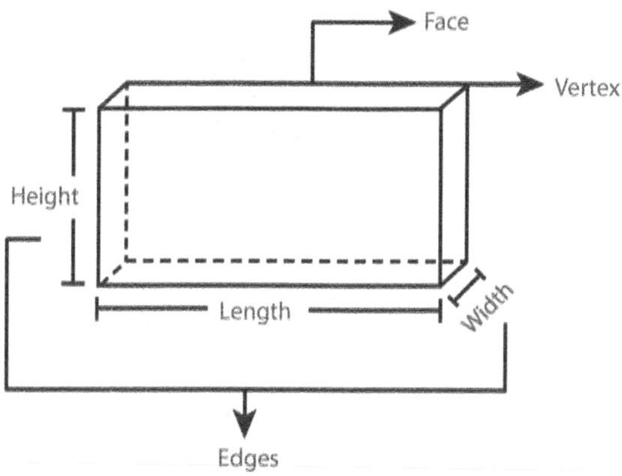

We see the world in 3-dimensional shapes. Some basic 3-dimensional shapes are cubes, rectangular prism, spheres, cones, and cylinders.

Shapes	Image	Real-life identical shapes	More examples
Cube			• Rubik's Cube • Ice and sugar cube • Building blocks • Boxes **Identify more**
Cuboid			• The lunch box • Cubicles • Shoebox • Book • Carton boxes **Identify more**

(Continued)

(*Continued*)

Shapes	**Image**	**Real-life identical shapes**	**More examples**
Cylinder			• Pipes • Battery • Water tanks • Gas cylinder **Identify more**
Cone	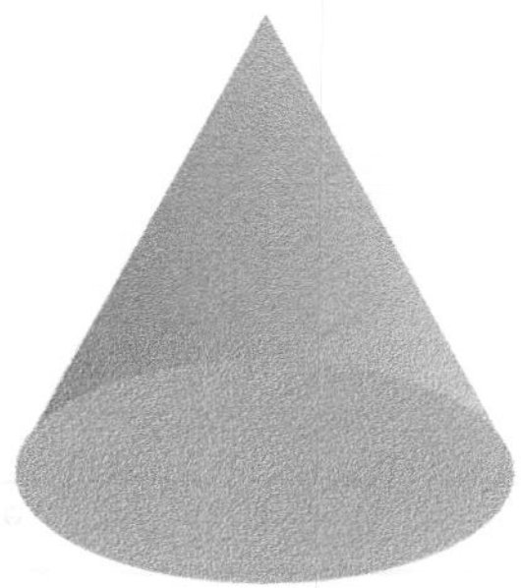		• Funnel • Christmas tree • Traffic cone • Waffle cone • Party hat **Identify more**

(*Continued*)

(*Continued*)

Shapes	**Image**	**Real-life identical shapes**	**More examples**
Sphere	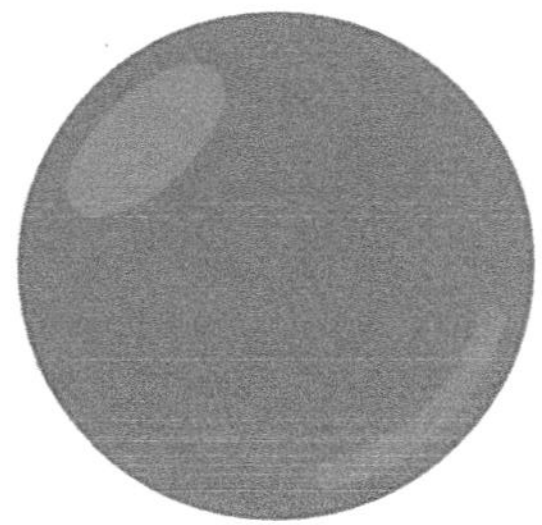		• Ball • Planets • Sun • Moon • Orange **Identify more**
Prism	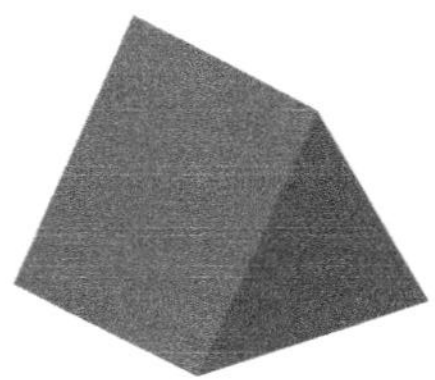		• Triangular roofs • Camping tents • Toblerone wrappers • Chocolate candy bars **Identify more**

(*Continued*)

(*Continued*)

Shapes	**Image**	**Real-life identical shapes**	**More examples**
Pyramid	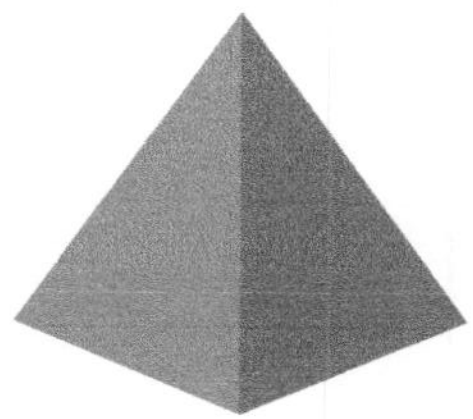		• Perfume bottles • Tents in army camps • The roof of a house **Identify more**

Formulas to find the Total Surface Area and the Volume

Name of the shape	Curved surface area	Total surface area	Volume
Cuboid	$2h(l + b)$	$2(lb + bh + hl)$	$l * b * h$
Cube	$4a^2$	$6a^2$	a^3
Cylinder	$2\pi rh$	$2\pi r(r + h)$	$\pi r^2 h$
Sphere	$4\pi r^2$	$4\pi r^2$	$4/3\pi\ r^3$
Cone	πrl	$\pi r(r + l)$	$1/3\pi\ r^2 h$
Hemisphere	$2\pi r^2$	$3\pi r^2$	$2/3\pi\ r^3$
Prizm			Area of the base × Length
Pyramid			1/3 (Area of the base × Height)

Let's understand these three-dimensional shapes in detail.

4.3.1 A Cube

A 3D shape made up of six equal squares is called a cube.

Classification of the cube: a cube has 6 square faces, 12 edges, and eight vertices(any three edges meet at one point called the vertex).

A cube is a 3D version of the shape square.
In a cube, length = height = width

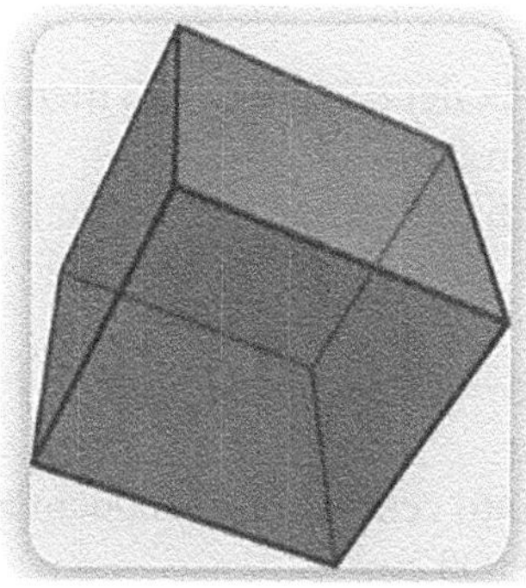

Formulas to calculate measurements of any cube
Note:

Lateral Surface Area (LSA)of a Cube	**4 × (length of the side)^2 units^2**
Total Surface Area(TSA) of a Cube	**6×(length of the side)^2 units^2**
The volume (V) of a cube	**(Side)^3 units^3.**

Volume can also be defined as the area of the base ×-height. Since the base is a square in a cube, the volume comes out to be the product of three sides of equal dimensions.

EXAMPLE 3: Find the lateral surface area and the total surface area of a cube of side 5 m.
SOLUTION: The length of the side of the cube is 5 m

$$\text{Lateral surface area is} = 4 \times [side]^2$$
$$= 4 \times 5^2$$
$$= 4 \times 25 = 100\ m^2$$

EXAMPLE 4: Find the cube volume with a side length of 8 cm.
SOLUTION: Let the side of cube = a = 8 cm

Volume of cube = $a^3 = 8^3 = 8 \times 8 \times 8 = 512$ cm^3

EXAMPLE 5: A cubical boiler can hold 1728 liters of water. Find the length of the side of the boiler in cm.
SOLUTION: We know that the capacity of any 3D shape is also considered as its volume.

Given that the capacity of the cubical boiler = is 1728 liters.
$\therefore$ Volume = 1728 liter
by converting littles into centimeters ($1000cm^3$=1 liter)
$\therefore$ Volume = a^3 = 1728000 cm^3
By taking the cube root of both sides
$\therefore$ a = 120 cm
Each side = 120 cm

EXAMPLE 6:- If the surface area of a plastic cubical-shaped box is 864 cm^2, find the amount (volume) of the plastic used to make the box.
SOLUTION: Given that the surface area of the cubical box is $864\ cm^2$

Let 'a' be the side of the cube.
Surface area = 6×(length of the side)^2 units^2
Therefore,

$$6a^2 = 864$$

$$a^2 = \frac{864}{6} = 144$$

$$\text{a} = 12 \text{ cm}$$

$$\text{The volume of plastic used} = a^3$$

$$\text{The volume of plastic used} = 12 \times 12 \times 12$$

$$\text{The volume of plastic used} = 1728\ cm^3$$

4.3.2 A Cuboid

A 3D shape of 3 sets of equal rectangles is called a cuboid. Classification of cuboid: a cuboid has 6 rectangle faces, 12 edges, and eight vertices(any three edges meet at one point called the vertex). A cuboid is a 3D version of a 2D shape rectangle. Length, height, and width (depth) are not equal in a cuboid in all three sets of rectangles.

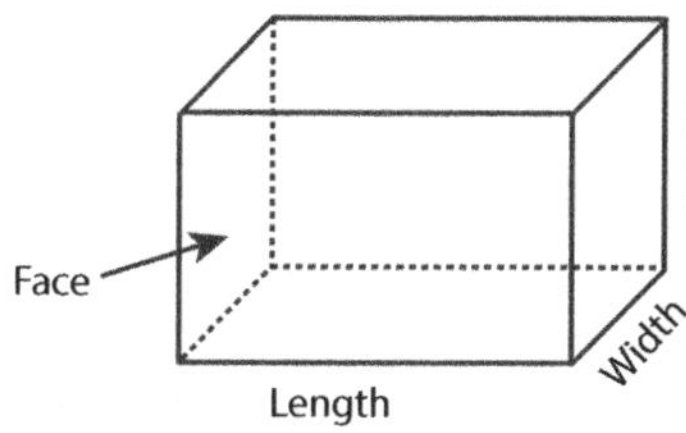

Formulas for the measurements of any cuboid

Lateral Surface Area (LSA)of a Cuboid	**2× height (length + width) = 2 × h (l + w) sq.units**
Total Surface Area (TSA) of a Cube	**2(length × width + length × height + height × width) = 2 × (l × w + w × h + h × l) sq.units**
The diagonal length of a Cuboid	$\sqrt{l^2 + w^2 + h^2}$ **Units.**
The volume (V) of a cube	**length × width × height = l × w × h cubic units**

Note: The Lateral Surface Area of a Cuboid is also called the area of the four walls.

As we find the area of any square or rectangle by counting the small squares within it, we can see the volume, too, by counting the number of small cubes in each layer of any cube or cuboid.

For example: we can find the given shape's volume by counting the small cubes.

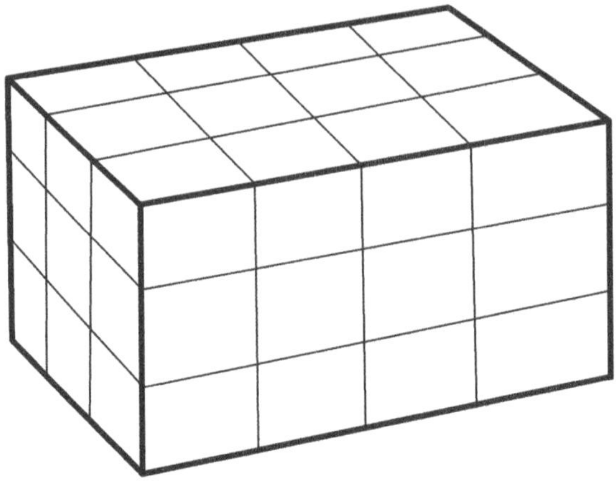

We can see that the given shape is a cuboid with three layers.
1st layer = 4 cubes horizontally × 3cubes vertically = 12 cubes in total
Total number of layers = 3
∴ Volume = 12 × 3 = 36 cubic. units

This method is proper when we are required to find the volume of ant irregular shape (a shape that does not have uniform dimensions).

EXAMPLE 7: Find the volume by counting the small cubes of the given irregular shape.

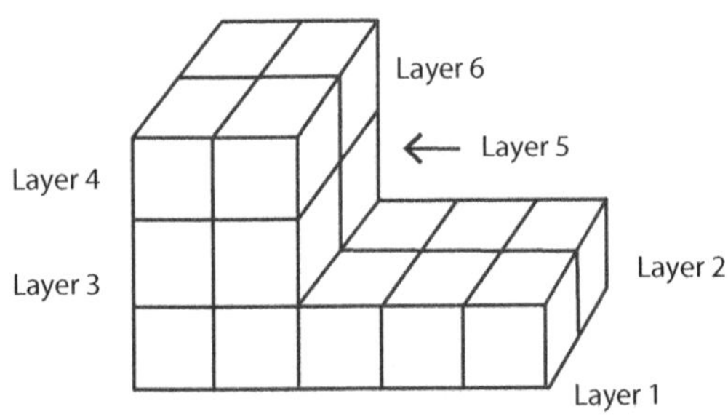

SOLUTION: There are many different ways to count the small cubes inside this shape; try this one
According to the given shape, count the number of small cubes in each layer separately, measuring $1cm^3$ each.

In the first layer, there are 5 small cubes. The second layer is at the back side, having the same number of cuboids, which is 5. The third, fourth, fifth, and sixth layers have 2 small squares each.

Hence the total Volume of the given 3D figure = 5 + 5 + 2 + 2 + 2 + 2 = 18 cm^3.

EXAMPLE 8: Find the given shape's volume by counting the small cubes. Consider the volume of each cube =1 cm^3.

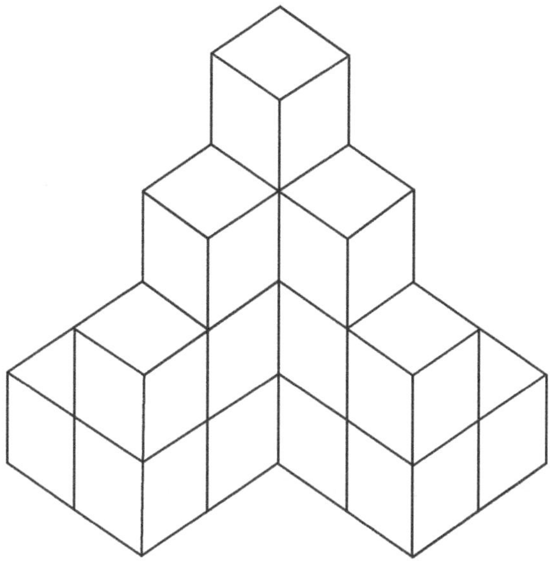

SOLUTION: identify the number of layers in shape.

We can see a total of 4 layers, counting the number of small cubes in each layer.

first layer (Topmost layer) = 1 cube

second layer = 3 cubes

third layer = 4 cubes

forth layer (bottom layer) = 6 cubes

Hence the Volume = 1 + 3 + 4 + 6 = 14 cm^3.

EXAMPLE 9: Find the volume of the given shape by counting the small cubes within it.

Consider the volume of each cube =1 cm^3.

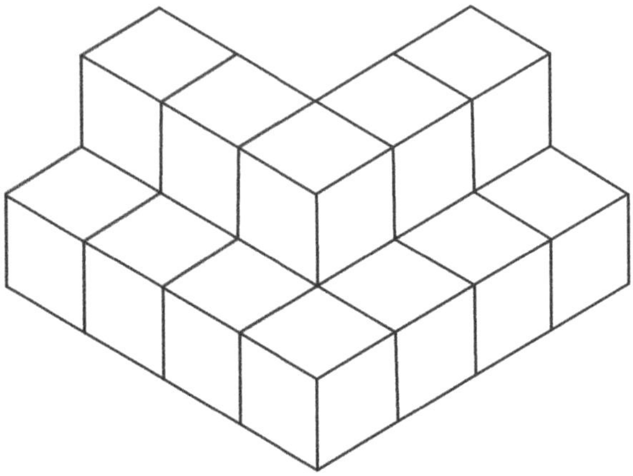

SOLUTION: We can see a total of 2 layers, counting the number of small cubes in each layer.

the first layer (Topmost layer) = 5 cubes

second layer = 12 cubes

Hence the Volume = 5 + 12 = 17 cm^3.

EXAMPLE 10: Find the volume of the given shape by counting the small cubes within it.

Consider the volume of each cube =1 cm^3.

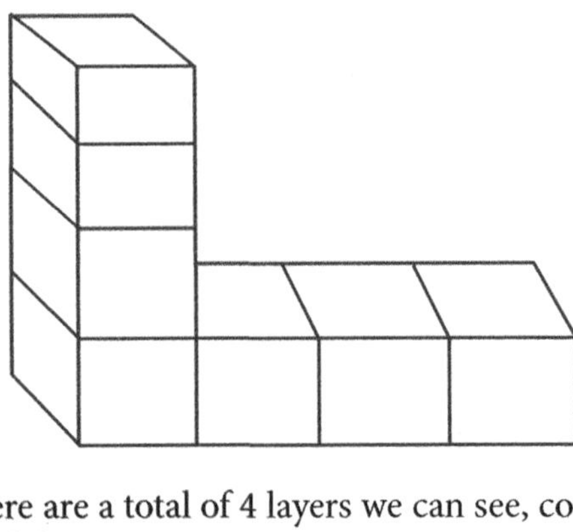

SOLUTION: There are a total of 4 layers we can see, count the number of small cubes in each layer.

first, second, and third layers (from the top) = 1cube each

forth layer = 4 cubes

Hence the Volume = 1 + 1 + 1 + 4 =7 cm^3.

EXAMPLE 11: Find the total and lateral surface area of a cuboid whose dimensions are 2 cm, 4 cm, and 6.5 cm.
SOLUTION: The three dimensions of the cuboid are

$$l = 2 \text{ cm}, \text{b} = 4 \text{ cm}, \text{h} = 6.5 \text{ cm}$$

$$\text{Total Surface area} = 2(l\text{b} + \text{bh} + \text{h}l)$$

$$= 2(2 \times 4 + 4 \times 6.5 + 6.5 \times 2)$$

$$= 2(8 + 26 + 13) = 2 \times 47 = 94\ cm^2$$

$$\text{Lateral Surface area} = 2 \times \text{h}(l + \text{b})$$

$$= 2 \times 6.5(2 + 4)$$

$$= 2 \times 6.5(6) = 78\ cm^2$$

EXAMPLE 12: An oil bottle is in the form of a cuboid which is 6 m in length, 5 m in width, and 4.5 m in height. Calculate how many liters this bottle can hold. (1 m^3=1000 liters).
SOLUTION: the capacity of the oil bottle =Volume of a cuboid = Length × width × Height

$=6 \times 5 \times 4.5 = 135\ m^3$

Since 1 m^3=1000 liters.

135 m^3 =1000 ×135 liters =135000 liters

EXAMPLE 13: What will be the height of a cuboidal box whose volume is 275 cm^3 and a base area of 5 cm^2?
SOLUTION: The volume of a cuboid = base area × height

$$\text{Height} = \frac{\text{volume of cuboid}}{\text{Base area}}$$

$$= \frac{275}{25} = 55 \text{ cm}$$

Therefore, the height of the cuboid is 55 cm.

EXAMPLE 14: The floor Area of a rectangular room is 80 m^2. Find its height if the volume is 370 m^3.
SOLUTION: The volume of the room = area of the base × height

$$\text{height} = \frac{\text{volume of room}}{\text{area of the base}} = \frac{370}{80}$$

$\therefore$ Height = 4.625m.

EXAMPLE 15: The dimensions of a room are 9 m × 8 m × 6 m. It has one door of dimensions 2.4 m × 1 m and two windows of dimensions 1.6 m x 1 m. Find the cost of whitewashing the walls at $5.80 per m^2.
SOLUTION: The walls are to be painted, so we removed the area of the door and two windows from it.

The area of the walls to be painted = area of the 4 walls – Area of 1 door – Area of 2 windows

Area of the walls to be painted = $2[\, h \times (l+b)] - 2.4 \times 1 - 2 \times 1.6 \times 1$

Area of the walls to be painted = $2 \times 6 \times \{9+8) - 2.4 \times 1 - 2 \times 1.6 \times 1$

Area of the walls to be painted = $204 - 2.4 - 3.2$

Area of the walls to be painted = $204 - 5.6$

Area of the walls to be painted = $198.4\ m^2$

Cost of whitewashing the wall = Area of the wall to be painted × cost per meter

Cost of whitewashing the wall =198.4 × 5.8 =1150.72 $

EXAMPLE 16: The swimming pool dimensions are 30 m × 10 m × 3 m. The depth of the pool is 3.5 meters. Find the volume of water in the pool in liters.
SOLUTION: Given that the length of the pool = 30 m

width = 10m, height = 3.5 m

Volume = $l \times w \times h = 30 \times 10 \times 3.5 = 1050\ m^3$

since 1 m^3= 1000 liter

$\therefore$ 1050 m^3 = 1050 × 1000 = 625000 liter.

EXAMPLE 17: A warehouse of dimensions 50 m × 30 m × 20 m is in the form of a cuboid. Find how many cubicles of volume 1.2 m^3 can be stored in it.
SOLUTION: Volume of one cubical box = 1.2 m^3

Volume of the warehouse = 50 × 30 × 20 = 30000 m^3

Number of boxes stored in the warehouse = Volume of the warehouse ÷ Volume of one box

= 50 × 30 × 20 ÷ 1.2 = 25,000 boxes

Hence, the number of cuboidal boxes that can be stored in the warehouse is 25,000.

EXAMPLE 18: Water is poured into a cuboidal tank at 80 liters per minute. If the volume of the pool is 240 m^3, how long will it take to fill the tank?
SOLUTION: Given that the rate of pouring water = 80 ltr. per minute

The volume of the pool = 240 m^3

Since, 1 m^3=1000 liters.

The volume of the pool = 240 000 l

Time taken to fill 80 L = 1 min

Time taken to fill 240 000 l = 240 000 × $\frac{1}{60}$ min

Time taken to fill 240 000 l = 4000 min

Time taken to fill 240 000 l = 66.7Hr approx.

EXAMPLE 19: In a particular area of society, there are 4000 residents. Each person's requirement is 150 liters of water every day. If the government built a water storage tank with the dimensions 20m ×15m × 6m storage, calculate the water in the tank will be sufficient for how many days?
SOLUTION: Per day total requirement of water = 4000 × 150 liter = 600000 liter

Volume of water in the tank $= 20 \times 15 \times 6 = 18000\, m^3$

Since, 1 m^3 = 1000 liter

Therefore, 18000 m^3 = 18000 000 liter

Number of days sufficient water in the $= \frac{18000000}{600000} = 30$ days

Hench, the water in the tank will be sufficient for 30 days.

EXAMPLE 20: If the dimensions of bricks are 25 cm×15 cm × 4 cm, calculate the required pieces of bricks to construct a wall of dimensions 24 m × 8m ×60 m if 10% of the wall is filled with cement.
SOLUTION: Volume of the one brick = 25 × 15 × 4 = 1500 cm^2
The volume of the material to build the wall.

$$= 24 \text{ m} \times 8 \text{ m} \times 60 \text{ m}$$

$$= 2400 \times 800 \times 60 \; cm^3 (\text{convert meter into centimeter: } 1 \text{ m} = 100 \text{ cm})$$

$$= 115200000 \; cm^3$$

Given that 10% of the wall is constructed with cement.

Hence, 10% of 115200000 $\frac{10}{100} \times 115200000 = 11520000$

Now, the volume of the material to build the wall with bricks =115200000-11520000= 1036800

Number of bricks required = $\frac{103680000}{1500}$ = 69120 bricks

EXAMPLE 21: A tank is rectangular. If the base area is 5500 cm^2, and the volume of water contained in it is 1.3 m^3. Find the level of water in the tank.
SOLUTION: Volume of rectangular tank = Area of base of tank × Depth of water

$$\text{Depth of water} = \frac{Volume}{Area}$$

$$= \frac{1.3m^3}{5500cm^3}$$

$$= \frac{1300000}{5500} (\text{convert m}^3 \text{ in to cm}^3,\ 1 \text{ m}^3 = 1000000\text{cm}^3)$$

$$= 236.36 \text{ cm} = 2.36 \text{ m}$$

EXAMPLE 22: If the length of the diagonal of a cuboid is 12 cm, and the sum of its length, width, and height is 21 cm. Calculate its surface area.
SOLUTION: According to the question

$$l + \mathrm{b} + \mathrm{h} = 21 \text{ cm}$$

$$\left[l + b + h\right]^2 = 21^2$$

Length of diagonal 12 cm

$$\sqrt{l^2 + b^2 + h^2} = 12$$

$$l^2 + b^2 + h^2 = 144$$

$$[l]^2 + [b]^2 + [h]^2 + 2[lb + bh + hl) = 411$$

$$144 + 2[lb + bh + hl) = 441$$

$$2\left[lb + bh + hl\right] = 441 - 144$$

$$2\left[lb + bh + hl\right] = 297\ cm^2$$

The surface area of the cuboid = 297 cm^2

4.3.3 Cylinder

A cylinder is a solid 3-dimensional shape with two circular bases and one curved side. The distance between the two circular bases is called the height of the cylinder. Soft drink cans, gas cylinders, etc., are real-life examples of cylinders.

Circular base cylinders are also known as right circular cylinders.

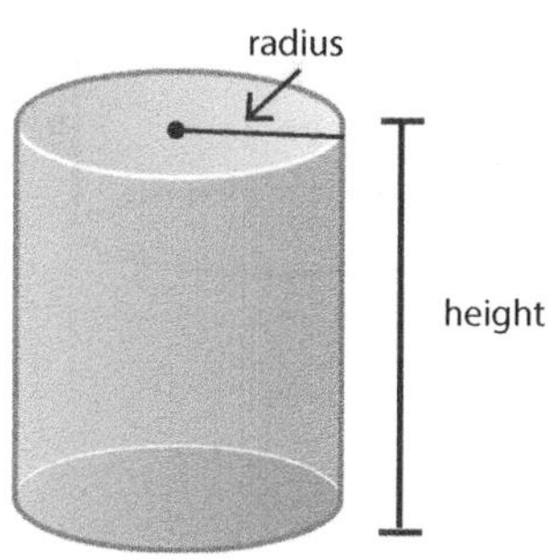

Surface Area and Volume

In simple words, we can say that the surface area of a cylinder means the area of the material used for making the cylindrical object. Observing a soft drink can, we can see the top and the bottom as two similar circles and one curved surface around the can. If we open or cut the can from the side and unroll it, we will see it is a rectangle. Try to check a water bottle and its label, which has a brand name or other details of the water; carefully remove the label from the bottle and look at the two-dimension shape which you get will be a rectangle. See the image below. A rectangle's length (L) is the circumference of the cylinder's base circle, and the can is the height of the cylinder.

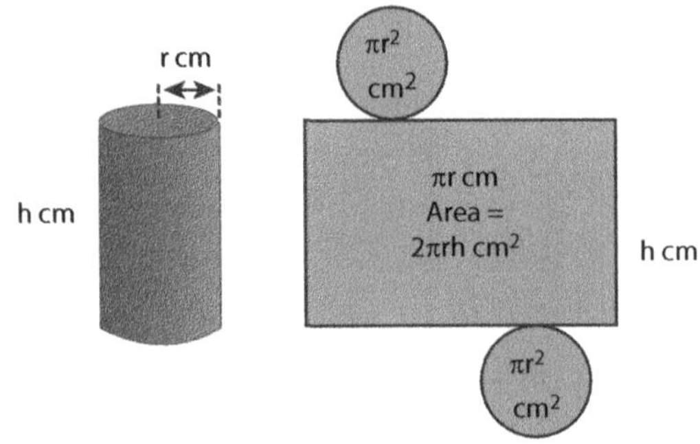

To find the total surface area of the cylinder, we add the areas of the two circles to the area of the rectangle.

The surface area of a cylinder with radius r and height h is.

$S = 2\pi r^2 + 2\pi rh$

Volume of the Cylinder

Volume of the Cylinder = (Area of the base) × height = $(\pi\, r^2) \times h = \pi\, r^2\, h$

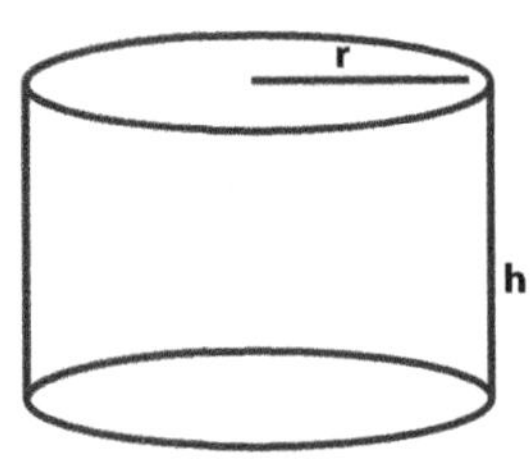

EXAMPLE 23: A cylindrical toy of length 20 m and radius 5 cm is made up of soft plastic. Find out how much plastic is used to make this toy? in a hot water heating system. Find the total radiating surface in the heating system.
SOLUTION: to calculate the plastic used to make this toy, we need to calculate its total surface area.

Given that the Length (h) = 20 m = 2000 cm
Radius (r) = 5 cm

$$\text{Curved surface area} = 2\pi rh$$
$$= 2\times\pi\times 5\times 2000 \text{ (Convert meter to centimeters, 1 m} = 100\text{cm)}$$
$$= 62831.85 \text{ cm}^2 = 628.31 \text{ m}^2$$

$$\text{Area of circles} = A = \pi r^2$$
$$= \pi\times 5\times 5 = 78.54 \text{ cm}^2 = 0.78\text{m}^2$$

$$\text{Total surface area} = \text{Curved surface area} + \text{area of the top and bottom circles.}$$
$$= 2\pi rh + \pi r^2 + \pi r^2$$
$$= 628.31 + 0.78 + 0.78$$
$$= 629.87 \text{ m or } 630\text{m (approximately)}$$

EXAMPLE 24: There are 14 support cylindrical pillars in a building. The radius of the base of each pillar is 20cm. And the height is 10m. How much concrete would be required to build all the pillars?
SOLUTION: Number of pillars =14

The radius of base (r) = 20 cm
Height of base (h)= 10 m = 1000 cm
Concrete required = 14 × Volume of one pillar
Concrete required = 14 × $\pi r^2 h$
Concrete required = 14 × $\frac{22}{7}$ × 20 × 20 × 1000

$$\text{Concrete required} = 14 \times \frac{22}{7} 20 \times 20 \times 1000$$

$$\text{Concrete required} = 1760000 \ cm^3$$
$$\text{Concrete required} = 17.6 \ m^3$$

EXAMPLE 25: A cylinder whose volume is 2.5 m^3 and has a diameter of 140 cm. Find the height. (use $\pi = 3.14$)
SOLUTION: Given that, the volume of cylinder = 2.5 m^3 and the diameter of base = 140 cm

$$\text{Hench the radius of the base} = \frac{140}{2} = 70\text{cm} = 0.7 \text{ m}$$

Let the height of the cylinder = h m

$$Volume = \pi r^2 h = 2.5$$

$$3.14 \times 0.7 \times 0.7 \times h = 2.5$$

$$h = \frac{2.5}{3.14 \times 0.7 \times 0.7} = 1.62 \text{ m}$$

EXAMPLE 26: The heights of cylinders A and B are in the ratio of 5:3, and their radius is in the ratio of 2:3. Find the ratio of the curved surface areas of both cylinders.

SOLUTION: The ratio of height $= \frac{h_1}{h_2} = \frac{5}{3}$

The ratio of radius $= \frac{r_1}{r_2} = \frac{2}{3}$

Let the ratio of Curved Surface Area $= \frac{c_1}{c_1} = \frac{2\pi r_1 h_1}{2\pi r_2 h_2} = \frac{2}{3} \times \frac{5}{3}$

The ratio of Curved Surface Area $= 10:9$

EXAMPLE 27: A painting roller is 9.5 cm long, and its radius is 3.5 cm. Covering one side wall of the room takes 550 complete revolutions. Determine the cost of painting the wall at the rate of 500 fils per m^2.(use $\pi = \frac{22}{7}$) (1000fils = 1 BHD, or 500 fils = 0.5 BHD)

SOLUTION: We can understand that the shape of the painting roller is a cylinder. For painting one full wall, the curved surface area of the painting roller is used.

Given that height of the painting roller = 9.5 cm, and radius = 3.5 cm

Curved surface area (C.S.A.) of a cylinder $= 2\pi rh$

$$= 2 \times \frac{22}{7} \times 3.5 \times 9.5 = 209 \text{ cm}^2$$

Total cost of painting a wall $= \frac{550 \times 209}{100 \times 100} \times 0.5$

$$= 5.748$$

$$= 6 \text{ BHD}(\text{app.})$$

4.3.4 Cone

A cone is a three-dimensional closed figure with a circular base connected to a vertex (or apex) point outside the plane of the base. In daily life, we can see so many conical shapes around us. One of the best examples is an ice cream cone. The height of a cone is the distance between its base and the vertex, and the height is always perpendicular to the base.

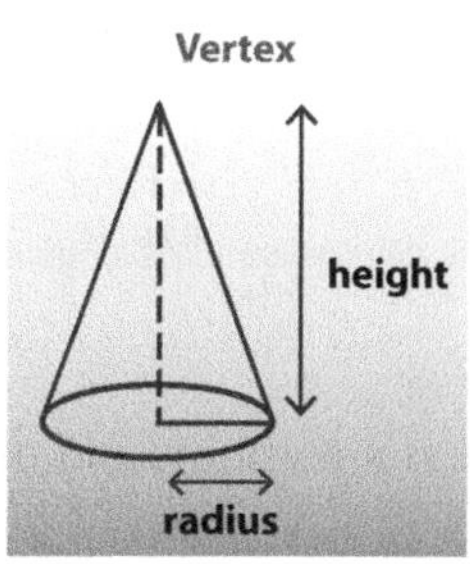

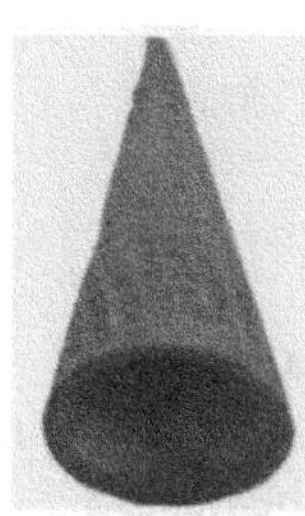

There are two types of cones, namely the right circular cone and oblique cone.

Right circular cone: If the segments joining the center of the circle base and vertex point are perpendicular to the base, the cone is a right circular cone.
Oblique circular cone: If the segment joining the center of the circle base and vertex point is not perpendicular to the base, the cone is called an oblique circular cone.

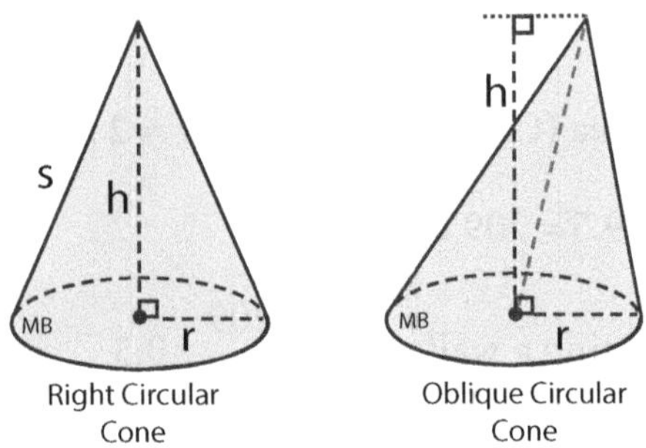

h = height; r = radius; s = slant height
some books use s as l (s=l)

When we cut any cone from its edge, we can see that the shape we got is a sector of the circle (Refer to the figure). The net of a cone includes a circle that gives the base, and a sector that gives the curved surface also called the lateral surface area.

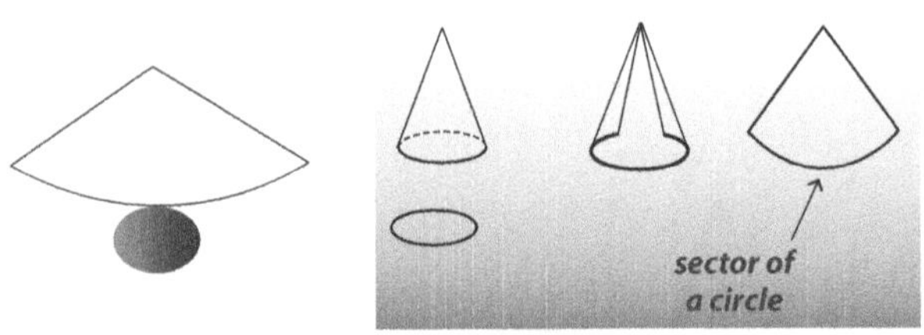

Surface area of a cone
The surface area of a closed right cone is a combination of the lateral area and the area of the base.

The formula for finding the total surface area of the cone = Area of the circle + area of the sector

T.S.A = $\pi r^2 + \pi rl = \pi r\,(r + L)$

The volume of the cone

There is a direct formula to find the volume of a cone.

$$\mathbf{V} = \frac{1}{3} \times \pi \times \mathbf{r}^2 \times \mathbf{h}$$

Where r refers to the radius of the circular base, and h is the height of the cone.

The volume of a cone is one-third the area of its base times its height.

But it is always interesting to know how we can drive the formula.

Let us try this, try to make a hollow cylinder and a hollow cone from paper with the same base radius and height (refer to the figure below), then, the cylinder volume will be three times the volume of the cone.

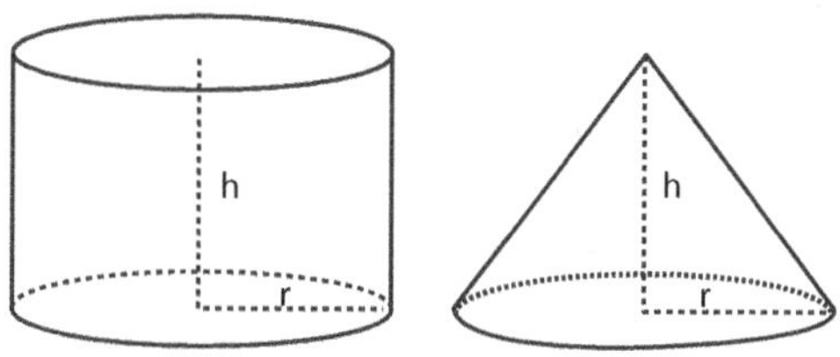

Since the volume of the cylinder is $V = \pi \times r^2 \times h$

Hence, the volume of the cone = $V = \frac{1}{3} \times \pi \times r^2 \times h$

This means it will fill the cylinder if we fill three identical cones with sand and pour them into a cylinder of the same height and radius.

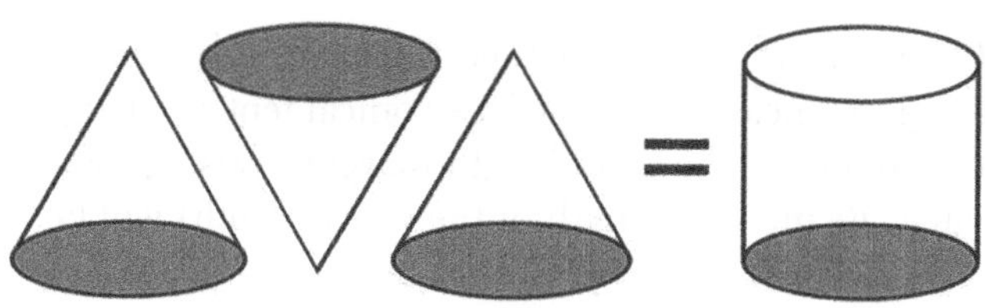

EXAMPLE 28: Find the Volume of the cone with the given information the radius =25 cm, and the height is 38 cm. (Use $\pi = 3.14$)

SOLUTION: Given that r = 21 cm and h = 28 cm

$$\text{Volume of cone} == \frac{1}{3} \times \pi \times r^2 \times h$$

$$V = \frac{1}{3}(3.14 \times 25 \times 25 \times 38)$$

$$V = \frac{1}{3} \times 74575$$

$$\therefore \text{Volume of a cone} = 24858.33 \text{ cm}^3$$

EXAMPLE 29: If the volume of a conical tank is 5430 cm^3, and its height is 85 cm. what will be the radius of its base?

SOLUTION: h = 85 cm and V = 5430 cm^3

$$\text{the volume of the cone} = \frac{1}{3} \times \pi \times r^2 \times h$$

$$\Rightarrow \qquad 5430 = \frac{1}{3} \times 3.14 \times r^2 \times 85$$

$$\Rightarrow \qquad 5430 = 88.97 \times r^2$$

$$\Rightarrow$$

$$r^2 = \frac{5430}{88.97} = 61.03$$

$$\therefore \qquad r = \sqrt{61.03} = 7.81 \text{ cm.}$$

EXAMPLE 30: Adam wanted to make a tent for her sister to play in. He is considering the measurements of the conical tent as a height of 8m and a base radius of 6m. If the extra length margins, wastage, and cutting are approximately 20cm, what length of the canvas(cloth) will be required to make the tent if its width is 3 m? (Use = 22/7)

SOLUTION: Area of the canvas required = curved surface area of the tent

Length of trampoline $= (l) = ?$

Breadth of trampoline(b) = 3 m

The radius of the tent $(\ ) = 6$ m

Height of the tent $(h) = 8$ m

$$l = \sqrt{h^2 + r^2} = \sqrt{36 + 64} = \sqrt{100}$$

$$l = 10 \text{ m}$$

$$l \times b = \pi \, rl$$

$$l \times 3 = \frac{22}{7} \times 6 \times 10$$

$$l = 62.86 \, m$$

Adding the wastage and margin of 20 cm
The total length of the trampoline required $62.86 + 0.20$
The total length of the trampoline required is 63.06 m (approx.)
and a height of 24cm

EXAMPLE 31: Sarah bought ten birthday caps in the shape of a right circular cone with a 7cm radius for her birthday. Calculate the number of paper sheets that will be required.
SOLUTION: Height of the cone(h) = 24 cm

The radius of the base $(r) = 7$ cm

Let the slant height of the cone be $= l$

$$l = \sqrt{h^2 + r^2}$$

$$l = \sqrt{24^2 + 7^2} = \sqrt{576 + 49}$$

$$l = \sqrt{625} = 25 \text{ cm}$$

The area of the sheet required for 10 conical caps $= 10 \, \pi \, rl$

$$= 10 \times \frac{22}{7} \times 7 \times 25 = 5500 \, cm^2$$

4.3.5 Sphere

A sphere is a three-dimensional closed figure. The volume of a sphere is the amount of space it occupies within it. The best daily life example of the sphere is a ball. The fixed distance between the points is the sphere's radius(r), and the fixed point is the sphere's center (C).

You can observe that if we rotate a circle horizontally or vertically, it will form a sphere. Hence, a sphere is obtained by rotating a two-dimensional object called a circle.

Surface Area of a Sphere

The surface area of a sphere is equal to the area of the entire face surrounding it. As a sphere has no flat surface. Therefore, Total surface area of a Sphere = Curved surface area of a sphere $=4 \times \pi \times r^2$

Volume of a Sphere

The volume of a sphere is the amount of space occupied by it. For a hollow sphere like a football, the volume can be viewed as the number of cubic units required to fill the sphere.

$$\text{The volume of the sphere} = \frac{4}{3} \times \pi \times r^3$$

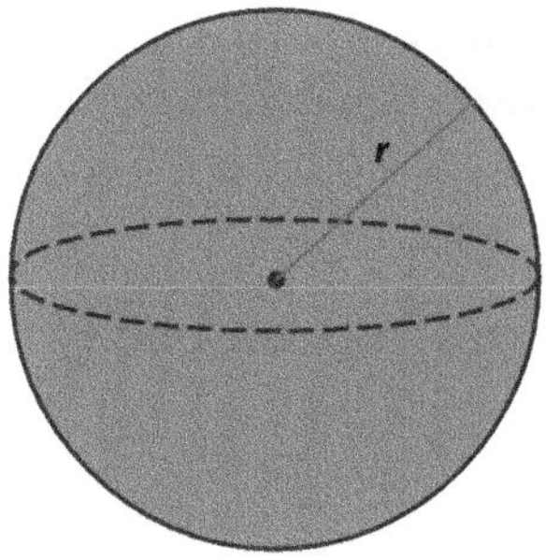

EXAMPLE 32: Calculate the volume of a sphere whose radius of 14.5 cm.

SOLUTION: Required volume

$$= \frac{4}{3} \times \pi \times r^3$$

$$= \frac{4}{3} \times \frac{22}{7} \times (14.5)^3 = 12775.19cm^3$$

4.3.6 Hemisphere

Divide the sphere into two parts, known as a hemisphere. Look around you in a few shapes. A soup bowl is the best example of a hollow hemisphere.

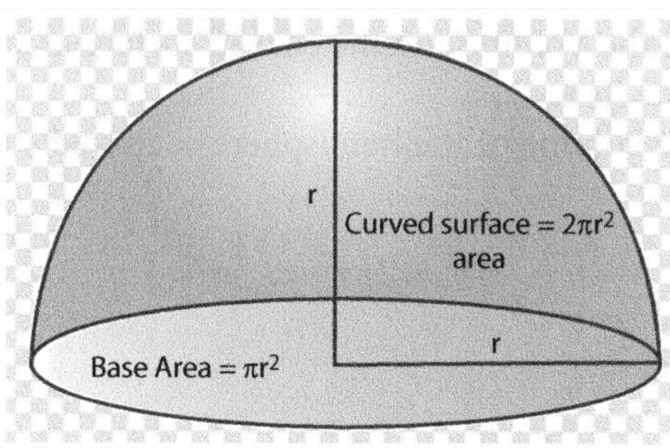

Surface Area of a Hemisphere

A hemisphere's surface area is equal to its surface's area, including the area of the circle at the top, as It has both curved and flat surfaces.

The curved Surface Area of a Hemisphere $= \frac{1}{2}$(curved surface area of the sphere) = units^2

$$= \frac{1}{2}\left(4 \times \pi \times r^2\right)$$

$$= 2 \times \pi \times r^2$$

Therefore, the total surface area of a hemisphere = Curved surface area of a hemisphere + area of the circle

$= (2\pi r^2) + (\pi r^2) = 3\pi r^2$

Volume of a Hemisphere

The volume of a hemisphere is the amount of space occupied by it. Same as a sphere, for a hollow hemisphere like a soup bowl, the volume can be viewed as the number of cubic units required to fill the hemisphere.

The volume of the hemisphere $= \frac{1}{2}$(volume of the sphere) $= \frac{2}{3} \times \pi \times r^3$

EXAMPLE 33: If the Volume and the curved surface area of a hemisphere are equal, calculate its radius.

SOLUTION: The volume of the hemisphere = curved surface area of the hemisphere

$$\frac{2}{3}\pi r^3 = 2\pi r^2 \Rightarrow \frac{1}{3} r = 1$$

radius = 3 units

EXAMPLE 34: A hemisphere soup bowl has a radius of 3.5 cm. Find out the quantity that this bowl can hold.

SOLUTION: Capacity of a hemispherical soup bowl

$$= \frac{2}{3}\pi r^3$$

$$= \frac{2}{3} \times \frac{22}{7} \times 3.5 \times 3.5 \times 3.5$$

$$= 89.83\ cm^3$$

4.3.7 Prism

A prism is a solid three-dimensional object whose front and back ends are similar in shape, equal and parallel rectilinear figures, and whose sides are parallelograms.

Any shape, regular or nonregular, can be a prism. In daily life, we see many shapes like our mobile phones, television sets, matchboxes, fish tanks, etc.

Prisms can be rectangular, triangular, or any other type.

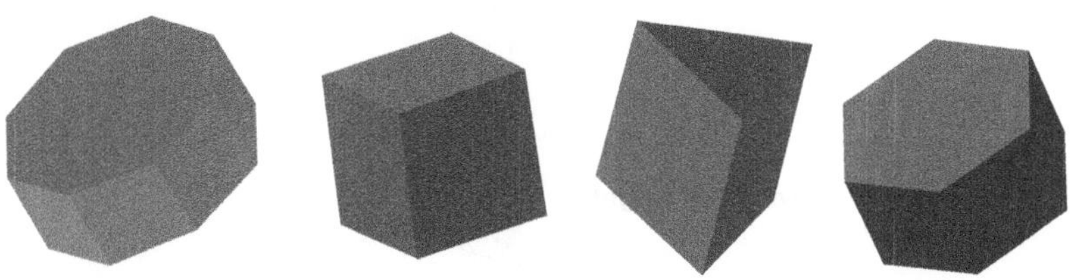

Surface Area of a prism

The surface area of Triangular Prisms 

A triangular prism consists triangular base or cross-section and three rectangular lateral faces.

Lateral surface area (L.S.A) = Perimeter of base × length = P × L

Total surface area (T.S.A) = 2 × Area of triangular base + L.S.A

The surface area of a rectangular Prisms 

The surface area of a rectangular prism is the same as the surface area of a cuboid.

Total surface area of a rectangular prism = 2(lb + bh + lh) square units where,

l = length of the rectangular prism
b = breadth of the rectangular prism
h = height of the rectangular prism
several sides in their base.

Volume of a Prism

As we know, to find the volume of any regular three-dimensional shape, we need to find out the base area or face area and multiply it by its length or depth.

Similarly, **The volume of Prism**: Base area of the shape × height/length/depth

EXAMPLE 35: A triangular prism having a base as an equilateral triangle of sides 6cm and 8 cm. Calculate its volume.

SOLUTION:

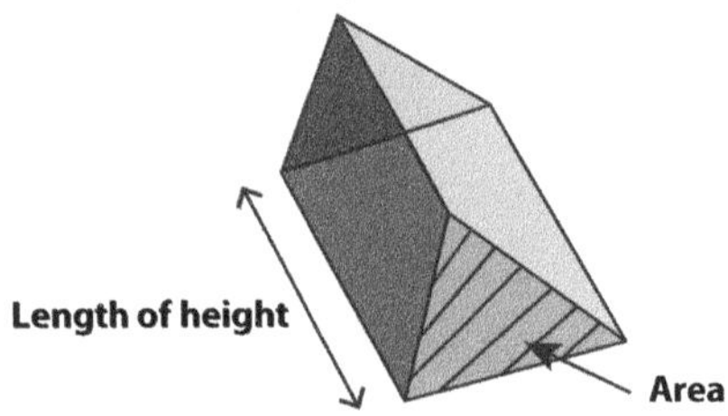

Given that the base of the triangular prism is equilateral (all three sides of the triangle are equal).

$$\therefore \text{Area of base} = \text{Area of the equilateral triangle} = \frac{\sqrt{3}}{4} \times (\text{side})^2$$

$$= \frac{\sqrt{3}}{4} \times 6^2 = 15.59 \text{ cm}^2$$

$$\text{Volume} = \text{Area of base} \times \text{height} = 15.59 \times 10$$

Hence the volume of the given prism = 155.88 cm^3

4.3.8 Pyramid

A Pyramid is a three-dimensional conical shape with a polygonal base (triangle, square, rectangle, hexagon, and so on) and vertical perpendicular height. It has a single vertex at the top connecting plane triangular faces.

The best example of a pyramid is "the Pyramids of Egypt."

We can observe so many different types of pyramids in our surroundings.

Any pyramid with a triangle base is called a triangular Pyramid, and with a rectangular base is called a rectangular pyramid, and so on.

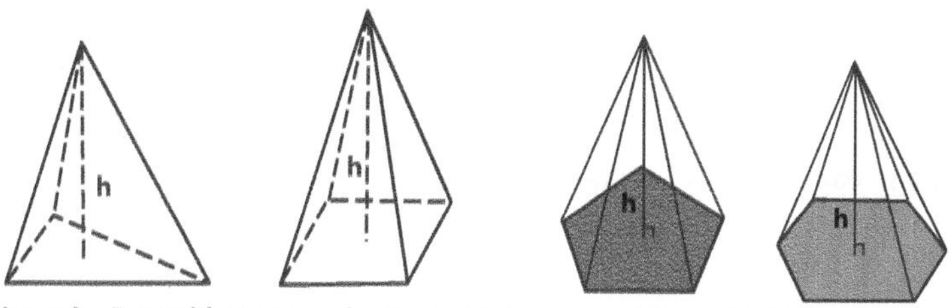

Triangular Pyramid Rectangular Pyramid Pentagonal Pyramid Hexagonal Pyramid

Surface Area of a Pyramid

Lateral Surface Area (LSA) = $\frac{\text{perimeter}}{2} \times$ slant height

Total surface area (T.S.A) = lateral surface area + area of the base

Volume of the Pyramid

The volume of the pyramid depends on the area of its base.

The volume of the Pyramid = $\frac{1}{3} \times$ base area $\times$ height

EXAMPLE 36: If the base of a triangular pyramid has a length of each side of 8 cm and a height of 12 cm. Determine its volume.

SOLUTION: Since the pyramid's base is triangular, all sides are equal. Hence the shape of the base is an equilateral triangle.

$$\text{Area of the equilateral triangular base} = \frac{\sqrt{3}}{4} \times (\text{side}) = \frac{\sqrt{3}}{4} \times (8)^2 = 27.71 \text{ cm}^2$$

Now, the volume of the pyramid = $\frac{1}{3}$ × base area × height = $\frac{1}{3}$ × 27.71 × 1 = 110.85 cm^3

EXAMPLE 37: If the base of a square pyramid has a length on each side of 3 cm and a height of 10 cm. calculate its volume.

SOLUTION: Area of square base = side × side = 3 × 3 = 9 cm^2

Now, the volume of the pyramid = $\frac{1}{3}$ × base area × height = $\frac{1}{3}$ × 9 × 10 = 30cm^3

EXAMPLE 38: If the base of a rectangular pyramid has the dimensions 7cm × 9 cm, and its height is 21cm. Find its volume.

SOLUTION: Area of rectangular base = length × width = 7 × 9 = 63 cm^2

Now, the volume of the pyramid = $\frac{1}{3}$ × base area × height = $\frac{1}{3}$ × 63 × 21 = 441cm^3

EXAMPLE 39: Find the surface area of the given shape.

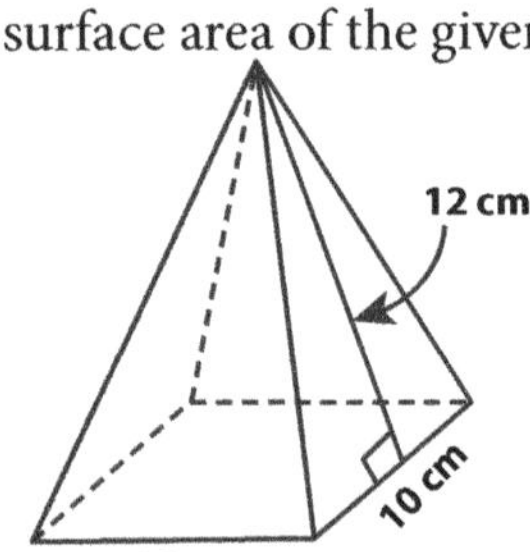

SOLUTION: Observe the figure given; the given figure is a square-based pyramid with base lengths of 10 cm and a slant height of 12 cm.

$$\text{Total surface area (TSA)} = \frac{\text{perimeter of the Base}}{2} \times \text{slant height} + \text{area of the base}$$

$$= \frac{10 \times 4}{2} \times 12 + (10 \times 10)$$

$$= 340 \text{ cm}^2$$

4.4 Compounded Shapes

When two or more shapes or figures join to make a new shape is known as a compounded shape.

Surface Area and Volume of any Compounded Shape

The surface area and volume of any three-dimensional compounded shape can be calculated by adding the curved surface areas or volume of each shape.

For example:

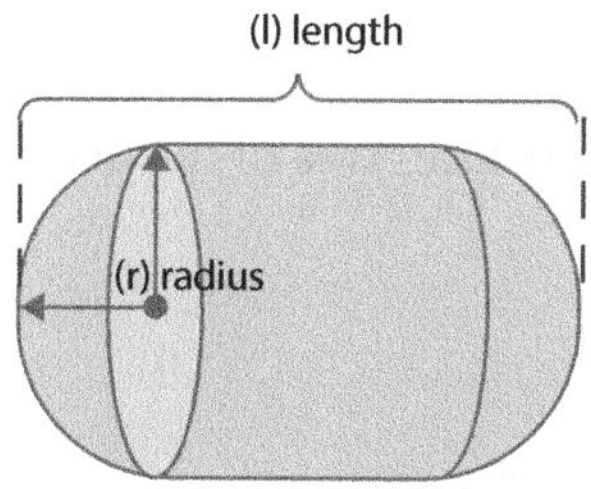

In the above shape, we can see two hemispheres (left end and right end) joining with the cylinder with radius r.

Hence the total surface area of the new shapes will be the sum of the curved surface areas of each shape.

The total Surface area of the new shape = CSA of two hemispheres + CSA of cylinder

Similarly, The volume of the new shape = volume of the two hemispheres (or one sphere) + volume of the cylinder

EXAMPLE 40: A party hall is made up of a combination of two solid three-dimensional shapes. The base of it is a cube of side 6 cm, and the top is a hemisphere. The base of the block is a cube, and the hemisphere of a diameter of 8 cm. How much paint is required to paint its full outer surface?

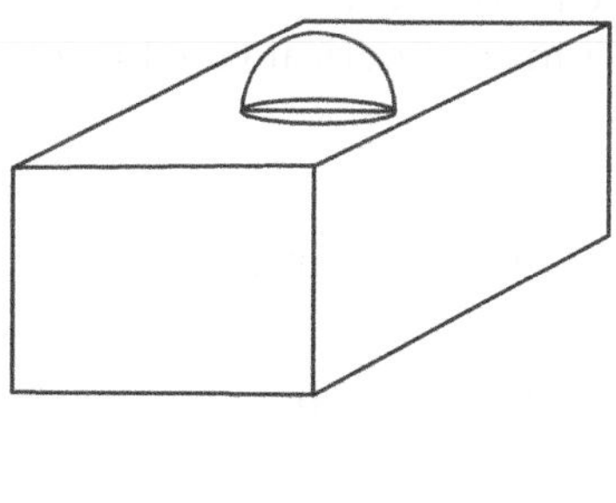

SOLUTION: Given that, the side of the square =6 cm

The diameter of the hemisphere =8cm

Hence, the radius of the hemisphere $= \frac{8}{2} = 4$ cm

The total surface area of the compounded shape = Total surface area of a cube + Curved Surface Area of the hemisphere − Area of a circle

$$= 6(side)^2 + 2\pi r^2 - \pi r^2$$

$$= 6(side)^2 + \pi r^2$$

$$= 6(6)^2 + \frac{22}{7} \times 4 \times 4 = 266.27 \text{ cm}^2$$

EXAMPLE 41: A water-full conical bowl has a radius of 5 cm and a height of 24 cm. This water is poured into a cylindrical bowl whose radius is 10 cm. Find out the height of the water in the cylindrical bowl. (Use the π button from your calculator).

SOLUTION: Given that, the radius (r_1) of the conical bowl =5cm

Height (h) of the conical bowl = 24cm

Radius (r_2) of the cylindrical bowl = 10cm

Let the height of the cylindrical bowl =H cm

To find the height of the water in the cylindrical bowl were required to use their volumes.

According to the scenario given,

The volume of water in a cylindrical vessel = volume of water in a conical vessel

$$2\pi r_2 H = \frac{1}{3} \pi r_1{}^2 h$$

$$2 r_2 H = \frac{1}{3} r_1{}^2 h$$

$$2 \times 10 \times H = \frac{1}{3} \times (5)^2 \times 24$$

$$20H = 200$$

$$H = \frac{200}{20} = 10\text{cm}$$

Therefore, the height of the water in the cylindrical bowl= 2 cm

Level 1: Exercise 4.1 Practice Questions

1. Calculate a cuboid's surface area and volume with the dimensions 15m×14m×7m.
2. Find the volume of a cube when its total surface area is 1350 cm^2.
3. The radius of a cylinder is 7 cm, and its total surface area of 968 cm^2. Find its height.
4. The surface area of a cube is 600 cm^2. Find the length of its side.
5. A small indoor greenhouse, including the base, is made entirely of glass panes held together with tape. It is 30cm long, 25 cm wide, and 25 cm high.
   a. What is the area of the glass used?
   b. How much tape is needed to join all 12 edges?
6. Find the height of the cuboid whose volume is 275 cm^3 and the area of the base is 25 cm^2.
7. Find the height of the cylinder whose volume is 1.54 m^3 and has a diameter of 140 cm.
8. Find each edge of the cube if the surface area of the cube is 96 cm^2.
9. The curved surface area of a right circular cylindrical vessel of height 14 cm is 176 cm^2. Find the diameter of the base of the vessel.
10. A metal pipe is 77 cm long. The inner diameter of a cross-section is 4 cm, and the outer diameter is 4.4 cm. Find the following:
    a. inner curved surface area
    b. outer curved surface area
    c. Total surface area
11. A round well's inner diameter is 3.5 meters. Find the inner curved surface area that needs to be plastered if it is 12 meters deep.
12. Find the curved surface area of a cone having a slant height of 50 cm and a base having a diameter of 28 cm.

Find the total surface area of a cone with a slant height of 21 cm and a base diameter of 24 cm.

13. The curved surface area of a cone is 4070 cm^2, and its radius is 35 cm. What is its slant height?
14. The curved surface area of a cone of radius 6 cm is 188.4 cm^2. Find its height.
15. The surface area of a sphere is 5544 cm^2. Find its diameter.
16. The radius of a cylinder is 14 cm, and its height is 40 cm. Find its:
    a. curved surface area
    b. total surface area
    c. volume
17. The curved surface area and pillar volumes are 264 m^2 and 396 m^3, respectively. Find the diameter and the height of the pillar.
18. A pond is 50 m long, 30 m wide, and 2 m deep. Find the capacity of the pond in cubic meters and liters.
19. A room's length, breadth, and height are 8, 6.5, and 3.5 m. What is the area of the four walls of the room?

Level 2: Exercise 4.2 Practice Questions

1. Three cubes whose edges are 3 cm, 4 cm, and 5 cm, respectively, are melted and recast into a single big cube. Find the surface area of the new cube formed.
2. The volume of a cuboid is 1536 m^3. Its length is 16 m, and its breadth and height ratio is 3: 2. Find the surface area of the cuboid.
3. A cone is given with a height of 15 cm. Suppose its volume is 1570 cm^3. Find the radius of its base. (Use $\pi = 3$).
4. The volume of a right circular conical flask is 9856 cm^3. If the diameter of the base is 28 cm, find
    a. height of the flask
    b. slant height of the flask
    c. curved surface area of the flask

5. Find the length of the longest rod that can be placed inside a room 12 m long, 8 m broad, and 9 m high.
6. A room's length, breadth, and height are 5, 4 and 3 m, respectively. Find the cost of whitewashing the room's walls and the ceiling at Rs. 8.50 per m^2.
7. A cardboard box 1.5 m long, 1.25 m wide, and 65 cm deep will be made open at the top. Ignoring the thickness of the cardboard used, find:
   a. area of the cardboard sheet required to make the box.
   b. the cost of the sheet for it, if the sheet costs Rs. 22/ m^2.
8. Sara wants to decorate her Christmas tree. She wants to place the tree on a wooden box covered with colored paper with a picture of Santa Claus on it. She needs to know the exact quantity of paper to buy for this purpose. If the box has length, breadth, and height of 80 cm, 40 cm, and 20 cm, respectively, how many square sheets of paper side 40 cm would be required?
9. The length and breadth of a hall are in the ratio 4:3 and its height is 5.5 m. The cost of decorating its walls (including doors and windows) at Rs. 6.60 per m^2 is Rs. 5082. Find the length and breadth of the room.
10. An iron pipe 20 cm long has an exterior diameter of 25 cm. The thickness of the pipe is 1 cm. Find the total surface area of the iron pipe.
11. A solid cylinder has a curved surface area of one-third of its total surface area. Suppose the total surface area is 462 cm^2. Find the radius and height of the cylinder.
12. You can make out the frame of a lampshade in the given figure. A piece of beautiful fabric is to be placed over it. The frame is 30 cm tall with a base diameter of 20 cm. For folding it over the top and bottom of the frame, a 2.5cm space must be left. To cover the lampshade, determine how much fabric is needed.

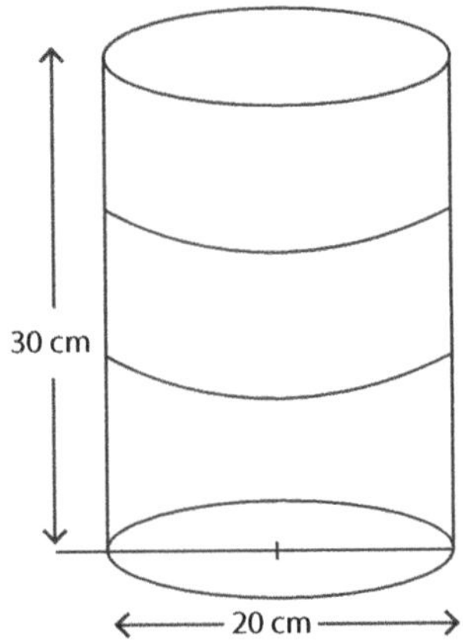

13. A cone's radius and slant height are in the ratio 4: 7. Find its radius if its curved surface area is 792 cm^2.
14. A conical tomb's slant height and base diameter are 25 m and 14 m, respectively. Find the cost of painting its curved surface area at Rs. 210 per 100 m^2.
15. A hemispherical bowl made of brass has an inner diameter of 7 cm. Find the cost of tin-plating it on the inside at Rs. 18 per 100 cm^2.
16. The base radii of two cones of the same height are in the ratio 3: 4. Find the ratio of their volumes.
17. A cone of height 12 cm has a curved surface area of 550 cm^2. Find its volume.
18. The radius and height of a right circular cone are in the ratio of 5: 12. If its volume is 314 cm^3, find the slant height and radius of the base of the cone. (Use $\pi = 31.4$).
19. Find the slant height and curved surface area of a cone whose volume is 12935 cm^3 and the base radius is 21cm.
20. A solid spherical ball of radius 3 cm is melted and then cast into smaller spherical balls, each of diameter 0.6cm. Find the number of small balls so obtained.

Level 3: Exercise 4.3 Practice Questions

1. The radius and height of a right circular cone are in the ratio of 5: 12. If its volume is 2512 cm^3, find the slant height and radius of the base of the cone. (use $\pi = 3.14$).

2. Find the percentage increase in the surface area of a cube when each side is doubled.
3. A rectangular box has dimensions of 2: 3: 4 and costs Rs. 650 more to cover it with a sheet of paper at the rate of Rs. 4 per m^2 than it does at the rate of Rs. 4.50 per m^2. Determine the box's measurements.
4. A rectangular chart of paper 88 cm × 50 cm is rolled along its length, forming a cylinder. Find the curved surface area of the cylinder so formed.
5. The total surface area of a hollow steel cylinder, open at both ends of an external radius of 8 cm and height of 10 cm, is 338 π cm^2. Find the thickness of the steel cylinder.
6. We know that the moon's diameter is approximately one-fourth of the earth's diameter. Find the ratio of their respective surface areas.
7. In the given figure, a right cylinder encloses a sphere of radius r. Find

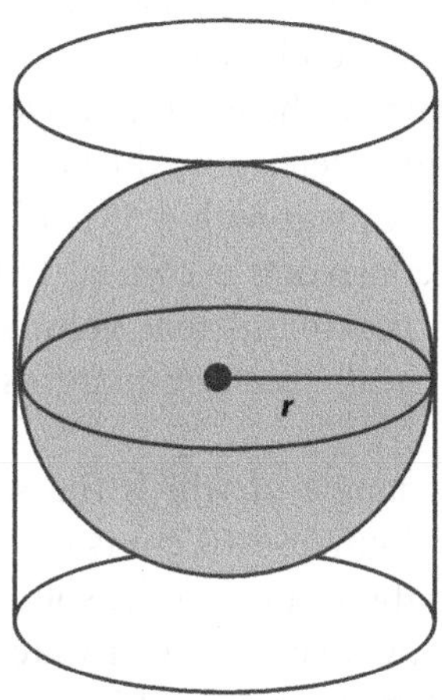

   a. The surface area of the sphere
   b. the curved surface area of the cylinder
   c. the ratio of the areas obtained in (a) and (b).
8. The areas of the three adjacent faces of a cuboid are 15 cm^2, 40 cm^2, and 24 cm^2. Find the volume of the cuboid.
9. The sum of a cuboid's length, breadth, and height is 21 cm, and the length of its diagonal is 12 cm. Find the surface area of the cuboid.

10. A rectangular tank is 120 m long and 75 m wide. At what speed per hour must water flow into it through a square pipe 20 cm wide so that the water rises by 2.4 m in 18 hours?
11. The sum of the height and radius of the base of a solid cylinder is 37 m. If the total surface area of the cylinder is 1628 m^2, find its volume.
12. A soft drink is available in two packs: a tin can with a rectangular base of a length of 5 cm and width of 4 cm, having a height of 15 cm, and a plastic cylindrical can with a circular base of diameter 7 cm and height 10 cm. Which container has greater capacity, and by how much?
13. The diameter of the cross-section of a wire is decreased by 5%. By how much percent should the length be increased so that the volume remains the same?
14. Water flows through a circular pipe, whose internal diameter is 2 cm, at 70 cm per second into a cylindrical tank. The radius of the base of the cylindrical tank is 40 cm. What will be the rise in the water level in half an hour?
15. A well was dug to a depth of 14 m with an inner radius of 4 m. To create a barrier, the earth removed from it is equally dispersed 3 m wide. Determine the height of this wall.
16. Sameer features a canvas with a 551 m^2 surface area. Using the canvas, she constructs a conical tent with a 7 m base radius. Find the maximum tent volume that can be built using it, assuming all stitching margins and cutting waste total about 1m^2.
17. A right-angled triangle of which the sides containing the right angle are 6.3 cm and 10 cm is made to turn round its longer side. Find the volume of the solid so formed.
18. In one minute, how many liters of water flow out of a pipe having an area of cross-section of 5 cm^2 if the speed of water flowing out from the pipe is 30 cm/sec?
19. A 3-cm spherical is lowered into a cylinder that is partially filled with water. The vessel's radius is 3 cm. Find how much the water level will rise if the sphere is fully submerged.
20. An ice cream cone has a hemispherical top of the same radius. If the height of the conical portion is 6 cm and the radius is 3.5 cm, find the ice cream volume in the cone.

4.5 Check Your Knowledge (MCQ)

1. $1m^3 =$ ________ cm^3
   a. 10,00,000
   b. 10,000,000
   c. 10,00,000,000
   d. 1,00,000

2. $1cm^3 =$ ________ dm^3
   a. 0.1
   b. 0.01
   c. 0.001
   d. 0.0001

3. A cube of side 4 cm is cut into 1 cm cubes. Then the ratio of the surface areas of the original cubes to the cut-out cubes is
   a. 1: 2
   b. 1 : 3
   c. 1: 4
   d. 1: 6

4. If the height of a cylinder decreases to $\frac{1}{4}$ the original height and the radius is doubled, then which of the following will be true?
   a. The curved surface area of the cylinder will be doubled.
   b. The curved surface area of the cylinder will remain unchanged.
   c. The curved surface area of the cylinder will be halved.
   d. The curved surface area will be $\frac{1}{4}$ of the original curved surface.

5. The dimensions of a warehouse are $40m \times 25m \times 10m$. If it is filled with cuboidal boxes, each of dimensions $2m \times 1.25m \times 1m$, find the total number of boxes that may be filled.
   a. 1800
   b. 2000
   c. 4000
   d. 8000

6. A metal sheet 27 cm long, 8 cm broad, and 1 cm thick is melted and reshaped into a cube. Then the side of the cube is
   a. 6 cm
   b. 8 cm
   c. 12 cm
   d. 24 cm

7. If the ratio of radii of two cylinders is 1: 2 and the ratio of their heights is 2:3. The ratio of their volumes will be
   a. 1:6
   b. 1:9
   c. 1:3
   d. 2:9

8. A cone is 8.4 cm high, and the radius of its base is 2.1 cm. It is melted and recast into a sphere. The radius of the sphere is
   a. 4.2
   b. 2.1
   c. 2.4
   d. 1.6

9. A Rectangular Sandbox is 5 m wide and 2 m long. How many cubic meters of sand is needed to fill the box to a depth of 10 cm?
   a. 2.5
   b. 1
   c. 5
   d. 3

10. What is the surface area of a cube of side 27 cm?
   a. 4374 cm^2
   b. 6080 cm^2
   c. 2298 cm^2
   d. 729 cm^2

11. If the length, breadth, and height of a cuboid are in the ratio 6: 5: 4 and the total surface area of the cuboid is 33300 cm^2. What is the breadth of the cuboid?
   a. 95 cm
   b. 25 cm
   c. 55 cm
   d. 75 cm

12. A beam is 9 m long, 40 cm wide, and 20 cm Deep. It is made up of iron that weighs 50 kg per cubic meter. What is the weight of the beam?
    a. 15 kg
    b. 39 kg
    c. 29 kg
    d. 36 kg

13. A Wooden Box of Dimensions 8 m × 7 m × 6 m is to carry Rectangular Boxes of dimensions 8 cm × 7 cm × 6 cm. What is the Maximum Number of Boxes that can be held in the Wooden box?
    a. 99,00,000
    b. 10,00,000
    c. 12,00,000
    d. 98,00,000

14. A Cube of side 6 cm is cut into some cubes, each side 2 cm. The number of Cubes so formed is
    a. 22
    b. 61
    c. 27
    d. 24

15. If the length of the diagonal of a cube is $4\sqrt{3}$ cm. What is the length of its edge?
    a. 4 cm
    b. 8 cm
    c. 6 cm
    d. 2 cm

16. What is the surface area of a cube if its volume is 2744 cm^3?
    a. 196 cm^2
    b. 1176 cm^2
    c. 784 cm^2
    d. 588 cm^2

17. If one coin is 1.5 cm in diameter and 0.2 cm thick, to be melted to form a Right Circular Cylinder of height 10 cm and diameter 4.5 cm. How many coins are required?
    a. 654
    b. 650
    c. 450
    d. 474

18. What is the total surface area of the cylinder if the ratio between the radius of the base and the height of a cylinder is 2 : 3 and its volume is 1617 cm^3?
    a. 462 cm^2
    b. 308 cm^2
    c. 707 cm^2
    d. 554 cm^2

19. What is the height of the cylinder if its curved surface area is 1760 m^2 and its base radius is 14 cm?
    a. 15 cm
    b. 5 cm
    c. 10 cm
    d. 20 cm

20. What is the total surface area of the cylinder having a diameter of 28 cm and a height of 20 cm?
    a. 2994 cm^2
    b. 2993 cm^2
    c. 2992 cm^2
    d. 2991 cm^2

21. If the surface area of a cube is numerically equal to its volume, then the edge of the cube is
    a. 8 Units
    b. 2 Units
    c. 6 Units
    d. 4 Units

22. To calculate the area of the road paved by a road roller, we need to find the ______ of the roller.
    a. Area
    b. Curved Surface Area
    c. Total Surface Area
    d. Volume

23. A ________, when rolled along its length or breadth, forms a Cylinder.
    a. Rectangle
    b. Square
    c. Cone
    d. Hexagon

24. If the curved surface area of a cylinder is numerically equal to its volume. What is the radius of the base?
    a. 1 Unit
    b. 2 Units
    c. 3 Units
    d. 4 Units

25. A classroom is 10 m long, 6.4 m broad, and 5m high. If each student is occupied 1.6 m^2 of the floor area, how many students can be accommodated in the classroom?
    a. 50
    b. 30
    c. 20
    d. 40

4.6 PBL/Experiment

On the weekend, you visited the mall with your little brother. He could see a lot of toys displayed, and he wanted them to buy a RUBIK's cube and strawberry ice cream. Observe the figures and answer the questions:

1. Calculate the volume of the RUBIK's cube if its side is 6 cm.
2. Calculate the total surface area of the RUBIK's cube if its side is 6 cm.
3. What is the curved surface area of the hemisphere (ice cream) if the base radius is 4cm?
4. Find the total volume of the Ice cream with its cone (Cone+Hemisphere).

5
Algebra and Graphs

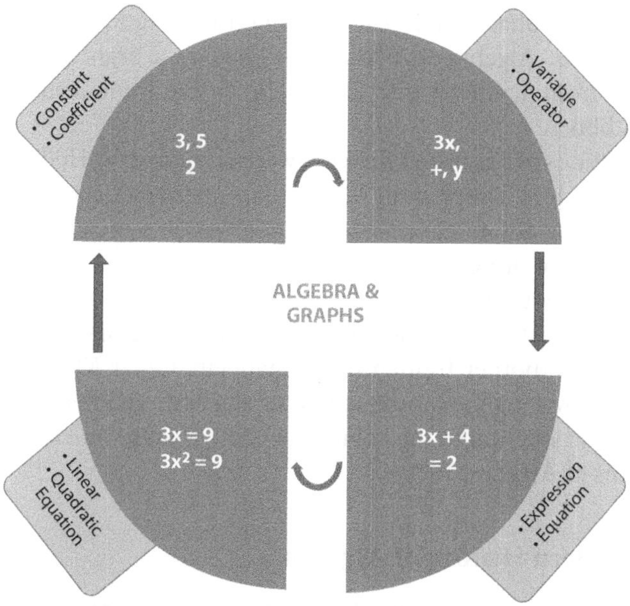

Objectives: This chapter

- *Explains how to identify patterns, represent real-world situations using expressions and equations, and simplify expressions using the order of operations.*
- *Introduce Graphs to solve linear, quadratic, and simultaneous equations.*

5.1 Introduction

"Algebra" is derived from the Arabic word written in the treatise of Medieval Persian Mathematician. The treatise provides a primary root of algebra and a systematic meaningful solution for the linear and quadratic equations.

Ritu Shrivastava, Ramakant Bhardwaj and Satyendra Narayan. Mathematics for Engineers, (171–248)

The late Middle English algebra style has come from Italian, Spanish, and medieval Latin, rooted in Arabic words.

5.1.1 Early History of Algebra

The ancient Babylonians invented formulas to resolve linear, quadratic, and indeterminate linear equations, which are the origins of algebra. Chinese, Greek, and Egyptian math used geometric techniques to determine these equations. Algebra was first developed as a branch of mathematics distinct from geometry and arithmetic by the Iranian mathematician Muammadibn.

The traditions of Egypt and Babylon were carried on by Hellenistic mathematicians such as Diophantus and Indian mathematician Brahmagupta (Brahmasphutasiddhanta, published in 628 AD), but on a higher level that incorporates zero or negative integers.

A debate about the entitlement of "the father of algebra" exists between the Greek mathematician Diophantus & Persian mathematician al-Khwarizmi. Another Iranian mathematician, Omar Khayyam, is credited with discovering the fundamentals of algebraic geometry and the general geometric solution to the cubic equation (Treatise on Demonstrations of Problems of Algebra, 1070).

5.1.2 Modern History of Algebra

Francois Viete created the modern version of algebra in the sixteenth century. Seki Kowa, a Japanese mathematician, and Gottfried Leibniz are credited with developing the concept of a determinant in the 17th century. In the 18th century, determinants and matrices were studied by Gabriel Cramer. Lagrange resolvents were first developed by Joseph-Louis Lagrange, who also researched permutations. Paolo Ruffini first created the theory of permutation groups. The development of abstract algebra took place in the 19th century. Axiomatic reasoning was created in algebra and mathematics by George Peacock. Augustus De Morgan created relationship algebra. Josiah Willard Gibbs developed three-dimensional vector algebra, and matrix algebra was developed by Arthur Cayley.

5.1.3 Algebra

Mathematics as a subject can be categorized into three significant branches: arithmetic, algebra, and geometry. algebra is considered one of the oldest elements in the history of mathematics. algebra describes the study of symbols, exponentials, known and unknown variables, and equations.

5.1.4 Algebra: What Is It?

A branch of mathematics known as "algebra" deals with manipulating symbols using arithmetic operators (addition, subtraction, multiplication, and division). The most popular symbols are alphabets, particularly x, y, z, p, and q, whose values change as they are altered and are hence called variables. Constants are values or numbers that remain the same as a calculation is performed. It's called a coefficient when a constant comes before a variable.

5.1.5 Why Should We Study Algebra?

Algebra makes it easier to translate mathematical expressions from actual calculating situations. For a mathematical expression to have meaning, it needs to include variables like x, y, and z and mathematical operations like addition, subtraction, multiplication, and division. Algebra is used in every area of mathematics, including coordinate geometry, calculus, and trigonometry. $2x + 4 = 8$ is a straightforward algebraic example.

5.2 Algebraic Expressions

Algebraic expressions have used the idea of expressing numbers using letters or alphabets. It does not specify their actual values. The basics of algebra teach people how to define an unknown value in terms of letters (e.g., x, y, z, etc.). These letters are known here as variables. An algebraic expression may also be a combination of both variables and constants. Any value placed before the variable (in other words, value multiplied by a variable) is a coefficient.

5.2.1 How are Expressions Formed?

The result of multiplying two variables, x, and y, by three, is $3xy$; in this case, the multiplication sign, known as the "term," can be omitted for convenience. $3xy$ is then added to another term, $6x$, which is mathematically expressed as $3xy + 6x$, an algebraic expression.

Hence, terms are added to form algebraic expressions.

For example: suppose an unknown number that is represented as x whose 3rd multiple is added to 15 is expressed mathematically as $3x + 15$.

Note: *The multiplicative things in a term are referred to as 'factors of a term.' In the above example, the first term 3x has 3 & x as factors.*

CONSTANT: Fixed Value

The value remains constant throughout the manipulation.

For example: 6, 5, 18 are constants.

VARIABLE: Symbol

The symbol carries value.

For example: x, y, x^2 are variables.

COEFFICIENT: Numerical factor of a variable

The coefficient is a constant value in front of a variable or multiplied by a variable.

For example: In $2x$, 2 is the coefficient of the variable x.

5.2.2 Arithmetic Operators

The operators (addition, subtraction, multiplication, and division) used in the algebraic expression are known as athematic operations. Usually, terms are combined by addition. Factors in terms are combined by multiplication, sometimes division.

For example: In $5x + 3y + 2$, $5x$, $3y$ & 2 are terms which are separated by + sign, an arithmetic operator.

Definition—Algebraic Expression

A mathematical statement that combines constants, variables, coefficients, and arithmetic operations is known as an "algebraic expression." The variables may be an exponent of a positive integer.

For example: $3x + 4$, $6x^2 + 3$

The generalized form of Algebraic Expression is $\mathbf{ax + b}$, where **a** is the coefficient of x, **a and b** are the constants, $\boldsymbol{x}$ is known as variable and '+' is called arithmetic operator.

For example:

(i) $6x + 8$ is an algebraic expression
Here 6 is the coefficient, x is the variable, 8 is the constant & and + is the arithmetic operator.

(ii) $7x^2 + 5x - 6$

Here 7 is the coefficient of x^2, 5 is the coefficient of x, 6 is the constant.

& +, - are arithmetic operators

(iii) $x^3 + 3x^2 - x + 1$

Here 1 is the coefficient of x^3, 3 is the coefficient of x^2, -1 is the coefficient of x, and 1 is the constant.

For example:

(i) Suppose a pen costs $ 6, and the service tax is $ 1 for each purchase. For Finding the cost of x pens, the algebraic expression used is $6x + 1$

(ii) A person can manufacture 16 items in one hour. The algebraic expression used to find the items manufactured by x persons is $16x$

(iii) The rent of hiring a car is $ 1500 per day, and the running cost is $ 10 per kilometer. The algebraic expression used to find the cost of hiring a car for a day that covers x km is $10x + 1500$

Definition—TERM of an Algebraic Expression

A single mathematical expression makes up the phrase. A term is either a single integer, also known as a "constant," "variable," or "coefficient and variable multiplied together," or it can be a set of constants. An arithmetic operator symbol (+), separating terms, is used.

Hence, term's are added to form an Expression.

For example: In, $8x^2 + 6x + y + 6$, $6x$, $8x^2$, 6, y are terms of the algebraic expression.

Note: If the '-' sign exists, we can consider it +(-1) and include (-1) as a factor of the following term.

For example: In, $5x + 8y - 3x$, the expression can be rearranged as $5x + 8y + (-3x)$, hence $5x$, $8y$ & $-3x$ are the terms of the algebraic expression.

***Note:** In 8x + 6x + 16, we can add the terms 8x and 6x, and the reduced algebraic equation is 14x + 16, which leads to the conclusion that terms with the same variables are referred to as like terms.*

Like Terms: Same Variable Terms
The terms whose **variables factors are the same,** which means **exponentially the same.**

For example: In $8x^2 + 5x + 3 + 6x + 3x^2$, $5x$ & $6x$ are like terms, $8x^2$ & $3x^2$ are like terms.

Unlike Terms: Different Variable Terms
The terms whose variable factors are different are called unlike terms.

For example: In $6x^2 + 3y^2 + 2xy + 2x + 3y + 4$, $6x^2$, $3y^2$, $2xy$, $2x$, $3y$ & 4 are unlike factors.

So, if we want to build a term, we have to multiply some constant referred to as a coefficient with variables that are referred to as factors. For example, the term $2xy$ is the product of 2, x & y, which are all the factors of the term.

Definition—Factors of a Term
The multiplicative things which cannot be factored further are known as factors.

For example: $8y^3$ can be written as $8 \times y \times y \times y$, and hence 8, y, y, y are the factors of the term.

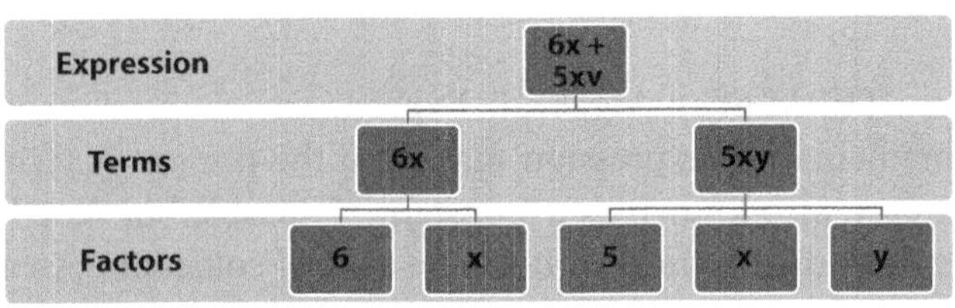

For example: look at equation $7y^3 + 5xy$; there are two terms, $7y^3$& $5xy$. In term $7y^3$, there are four factors 7, y, y & y. In the $5xy$ term, there are three factors 5, x & y.

Note: The equation is classified into different types depending on the number of terms as follows:

MONOMIAL: Single Term
An algebraic expression with only one term is called a monomial.
For example: 6, x^5, 8x, 8y/3 are Monomials.

BINOMIAL: Two Terms
An algebraic expression with two terms is called a binomial.
For example: $6x + 4$, $8y - 3$,

TRINOMIAL: Three Terms

A three-term algebraic expression is known as a trinomial.
For example: $2x + 6y + 3$, $5x^2 + x + 3$

POLYNOMIAL: One or More Terms

An algebraic expression with one or more terms is called a polynomial.
For example: $2x$, $2x + 3$, $5x + 3y + 2$

More examples:

(i) In $6x = 3$, the number of terms in the L.H.S. of the equation is one. Hence, it's a Monomial.
(ii) In $4x + 3 = 7$, there are two terms $4x$ & 3 on the left-hand side of the equation. Hence, it's a Binomial.
(iii) In $8x^2 + 6x + 4x + 9 = 0$, After reduction, $8x^2 + 10x + 9 = 0$, there are three terms. Hence, the equation is a Trinomial. All the above examples are polynomials.

5.3 Transposition of Formulae

Formulae are used frequently in almost all engineering, commerce, and medical fields to relate a physical quantity to determine unknown quantities. Most of the physics laws are described using formulae. For example, you may have already seen Ohm's law, V = IR, or Newton's second law of motion, F = m×a.

Example: VOLTAGE = CURRENT × RESISTANCE

$$R = \frac{V}{I}$$

Here we rearranged each of the quantities such that we could determine the resistance values.

In formulas with numerous variables, we must move all the known variables to the right-hand side of the equal sign and all the unknown variables to the left.

For example: When the volume and radii of a cylinder are known, we must determine its height.

The volume of the cylinder $= \pi r^2 h$

$$\therefore \textbf{Height } h = \frac{V}{r^2 h} \quad \textbf{(Transposing)}$$

For example: The cone's slant height, for which the height and radius are specified

By Pythagoras' theorem,

$$l^2 = r^2 + h^2$$
$$\therefore l = \pm\sqrt{r^2 + h^2}$$

$l = \sqrt{r^2 + h^2}$ Since l is a positive quantity

For example: the sphere's radius when the volume is known.

The volume of sphere $V = \frac{4}{3}\pi r^3$ and $r = 3\sqrt{\frac{3V}{4\pi}}$

Level 1: Exercise 5.1 Practice Questions

1. How many variables exist in the algebraic expression $x^2 + 4x + 3$?
2. Find the coefficient of x in the algebraic expression $6x + 4x$.
3. How many terms are included in the phrase $8x^3 + 6x^2 + 9x + 3$?
4. Extract the constant in the algebraic expression $9x^3 + 8x^2 + 6x + 3$.
5. Match the coefficients of x for the following expression.
   a. $5x^2 + 8x + 3$
   b. 8y
   c. $6x^2 - 7x + 1$
   d. -7
   e. $8xy + 6y + 5$
   f. 8
6. Find which of the following pairs are like terms.
   a. $12x$, $8y$
   b. $13xy$, $15yx$
   c. $8x$, $-6x$
   d. $8x^2y$, $6yx^2$

7. Form an algebraic expression with the statement
   a. "Eight times a variable x is added to 13"?
   b. "Sum of the product of x with y and x with 18?
   c. "Sum of the squares of x and 3"?
8. Identify the terms which contain x and find the coefficient of x in the expression $9x^2 + 6xy + 8$.
9. State which pair is unlike terms.
   a. $5x, 8xy$
   b. $8x^3, -6x^3$
   c. 2, 18
10. Identify which of the following are algebraic expressions.
    a. $8x + 7y + 3$
    b. $8\sqrt{x} + 3y + 15$
    c. $8y^3 + 6^{1/3}$
11. Transpose of the following formula and equations and obtain the variable's value in the bracket.
    a. $S = \frac{d}{t}$, (t)
    b. $v = u + at$, (t)
    c. $v^2 - u^2 = 2as$, (a)
    d. $s = ut + \frac{1}{2}at^2$, (a)
    e. $y(x + 1) = 3x + 1$, (x)
    f. $\frac{y}{y+x} + 6 = x$, (y)
    g. $\frac{t}{t-s} = 3s$, (t)
    h. $a(a - b) = a - 1$, (b)
    i. $m = a + \frac{1}{n}$, (n)
    j. $k = \frac{mv^2}{2}$, (v)

Level 2: Exercise 5.2 Practice Questions

1. Write examples of monomial, binomial, and trinomial expression equations.
2. Write an algebraic expression whose coefficients are 3, 4, and -6.

3. Write an algebraic expression with six terms; each of the coefficients is 7, 12, 14, 21, 28, 29.
4. Write an algebraic expression with five terms, including a constant term.
5. Write an algebraic expression with i) one variable, ii) 3 coefficients, iii) degree 3.
6. Write an example of like and unlike terms
7. Simplify like and unlike terms $5xy, 9x, 12xy, 13x^2, 5x, 14x^2, 12x^2y$ & $19xy^2, 20x^2y, 19xy^2$
8. Write a trinomial algebraic expression whose coefficients are 13, 14, and 17.
9. Give examples of Monomial, binomial, and trinomial expressions.
10. Identify variables, coefficients, and constants in the expression $\frac{a^2}{4}-\frac{a}{3}+\frac{b}{7}+12$
11. Transpose of the following formulae and equations to the variable mentioned in the brackets.

    a. $\mathrm{p}=\frac{P_o}{1-x^2}$ (x)

    b. $\mathrm{V}=\frac{V_o}{\sqrt{r^2-1}}$ (r)

    c. $\frac{1}{f}=\frac{1}{u}+\frac{1}{v}$, (v)

    d. $\mathrm{E}=\frac{1}{2}mv^2+\mathrm{mgh}$, (v)

    e. $v=\frac{1}{3}\pi r^2 h$, (h)

    f. $\mathrm{A}=2\pi r^2+2\pi rh$, (h)

    g. $\mathrm{E}=\frac{1}{2}mv^2+mgh$, (h)

    h. $\frac{s^2}{2t-s}+12=5\mathrm{t}$, (t)

    i. $\mathrm{y}=x+\frac{1}{1-a}$, (a)

    j. $\mathrm{m}=\mathrm{k}\sqrt{a(1-2x)}$, (x)

    k. y (3x +1) + y(3x-1) = 5x+3, (x)

    l. $\frac{1}{2}y\left(x-\frac{1}{3}\right)-\frac{1}{3}y\left(x+\frac{1}{3}\right)=\frac{x}{2}+4,(y)$

m. $T = 2\pi\sqrt{\frac{l}{g}}, (g)$

n. $F = G\frac{m_1 m_2}{r^2}, (r)$

5.4 Algebraic Fractions

Definition—Algebraic Fractions

A fraction with algebraic expressions referred to as 'terms' on both the numerator referred to as 'dividend' and the denominator referred to as 'divisor' provided the value of the algebraic expression in the denominator should not be zero. For example: $\frac{5x+4}{6x+5}, \frac{\sqrt{3x+2}}{3x+1}$

Depending on the terms of the algebraic fractions, there are two types.

Rational Functions

The terms of the algebraic fractions are polynomials (whose variables have positive exponent)

For example: $\frac{5x+4}{6x+5}$

Irrational Functions

The terms of the algebraic fractions are not polynomial (whose variables have fractional exponent)

For example: $\frac{x^{1/2}-3}{x^{1/3}-2}$

EXAMPLE 1: Find the reduced fraction of the algebraic fraction $\frac{4x+4}{12x+4}$

$$\textbf{SOLUTION:} \ \frac{4x+4}{12x+4} = \frac{4(x+1)}{4(3x+1)} = \frac{x+1}{3x+1}$$

5.4.1 How Can We Calculate Two Algebraic Fractions?

We can calculate two algebraic fractions by using the same laws of arithmetic fractions. The denominators of the two algebraic fractions must be the same for the addition and subtraction of two algebraic fractions. Otherwise, we have to make the denominators the same.

Addition and Subtraction of two algebraic fractions with the same denominator

Change the same denominator but add or remove the two numerators.

For example: $\dfrac{5x+4}{5}+\dfrac{2x+3}{5}=\dfrac{(5x+4)+(2x+3)}{5}=\dfrac{7x+7}{5}$

With different denominators

Determine the Least Common Denominator, or the least common multiple, between the two denominators (LCD). Use the LCD as the denominator to transform the two fractions into equivalent fractions. Place the two numerators above the LCD and multiply or divide them.

For example:

$$\frac{5x+4}{3}+\frac{2x+3}{5}=\frac{5x+4}{3}\times\frac{5}{5}+\frac{2x+3}{5}\times\frac{3}{3}=\frac{5(5x+4)+3(2x+3)}{15}$$

$$=\frac{25x+20+6x+9}{15}=\frac{31x+29}{15}$$

Multiplication of two algebraic fractions

We have to multiply the numerators together and denominators together.

For example: $\dfrac{5x+4}{2x+1}\times\dfrac{2x+3}{5x+2}=\dfrac{(5x+4)(2x+3)}{(2x+1)(5x+2)}$

Division of two algebraic fractions

Invert the second fraction by exchanging the numerator and denominator. Then change the division operator to multiplication.

For example: $\dfrac{2x+4}{2x+1}\div\dfrac{x+2}{x+3}=\dfrac{2(x+2)}{2x+1}\times\dfrac{x+3}{x+2}=\dfrac{2(x+3)}{(2x+1)}$

More examples:

(i) Add: $\dfrac{6}{5x}+\dfrac{8}{5x}=\dfrac{6+8}{5x}=\dfrac{14}{5x}$

(ii) Add: $\dfrac{5x}{x+3}+\dfrac{8}{x+3}=\dfrac{5x+8}{x+3}$

(iii) Add: $\dfrac{2x}{9}+\dfrac{3x}{3}=\dfrac{2x}{9}+\dfrac{3x}{3}\times\dfrac{3}{3}=\dfrac{2x+9x}{9}=\dfrac{11x}{9}$

(iv) Add: $\dfrac{3x}{x+2}+\dfrac{4x+1}{x+3}=\dfrac{3x(x+3)+(x+2)(4x+1)}{(x+2)(x+3)}$

$$=\frac{3x^2+9x+4x^2+x+8x+2}{x^2+2x+3x+6}=\frac{7x^2+18x+2}{x^2+5x+6}$$

EXAMPLE 2:

(i) Subtract $\frac{2x}{x+1} - \frac{1}{x+1}$

(ii) Subtract $\frac{x}{x+1} - \frac{2x+1}{x}$

SOLUTION: (i) $\frac{2x-1}{x+1}$

SOLUTION: (ii) $\frac{x^2-2x-1}{x(x+1)}$

EXAMPLE 3:

(i) Multiply $\frac{2x}{x}\, by \frac{(4x+1)}{x+1}$

(ii) Divide $\frac{2x}{x+1}\, by \frac{x+2}{x+1}$

SOLUTION: (i) $= \frac{2x(4x+1)}{x(x+1)}$

SOLUTION: (ii) $= \frac{2x}{x+1} \times \frac{x+1}{x+2} = \frac{2}{x+2}$

Level 1: Exercise 5.3 Practice Questions (Algebraic Fractions)

Simplify:

1. $\frac{p+q}{3} - \frac{3p}{6}$
2. $\frac{3}{5m} - \frac{n-1}{6mn}$
3. $\frac{2}{m+1} - \frac{3}{m-n}$
4. $\frac{y}{y+1} - \frac{2y}{y^{-3}}$
5. $\frac{6}{2x} + \frac{10}{5x}$
6. $\frac{m}{16} \div \frac{5m^2}{32}$

7. $\frac{8(x-3)}{24} \times \frac{12(x+3)}{4(x^2-9)}$

8. $Add \frac{x}{2} + \frac{3x}{2}$

9. $Add: \frac{x}{3} + \frac{x}{4}$

10. $Add: \frac{y}{4} + \frac{y}{2} + \frac{y}{8}$

Level 2: Exercise 5.4 Practice Questions (Algebraic Fractions)

Evaluate the Algebraic Fractions:

1. Multiply: $\frac{y}{2} by \frac{y}{3}$
2. Multiply: $\frac{y^2}{4} by \frac{y^3}{3}$
3. Multiply: $\frac{y+1}{y} by \frac{y^5}{y^{-1}}$
4. Multiply: $\frac{y^2-9}{y^{-3}} \times \frac{y+3}{3}$
5. Simplify: $\dfrac{\frac{m+2}{3} - \frac{m-2}{6}}{\frac{m-4}{4}}$
6. Simplify: $\frac{(m+2)^2-(m-3)^2}{6(m-1)}$
7. Simplify: $\frac{6x+3}{9} \div \frac{2x-1}{18}$
8. Simplify: $\frac{3x^3-9x^2}{3} \div \frac{4x^2-16}{4}$
9. Simplify: $\frac{9x-3}{3} + \frac{2x+3}{6} + \frac{5x-3}{12}$
10. Simplify: $\dfrac{\frac{9x}{7}}{\frac{4x}{14}} + \dfrac{\frac{16x}{7}}{\frac{40x}{5}}$

Level 3: Exercise 5.5 Practice Questions (Algebraic Fractions)

Simplify

1. $\frac{-9}{27}xyz\left(\frac{4}{7}x^2yz-\frac{3}{2}xyz^2\right)$
2. $0.5y\left(\frac{0.2}{0.03}xy^2-\frac{0.3}{0.06}x^2y\right)$
3. $\left[\frac{\left(\frac{4}{3}mnp\times\frac{3}{4}mnp\right)}{\left(5m^2n^2p^2\right)^2}\right]^{-\frac{1}{2}}$
4. $x^2+2x^2+3x^2+4x^2+\ldots+100x^2$
5. *Find the value of* $\left(\frac{1}{2}m^2+\frac{3}{4}n^2-\frac{8}{7}p^2\right)^2$ *for* $m=2, n=1, p=3$

Level 1: Exercise 5.6 Practice Questions (Addition and Subtraction)

1. Add the following
   a. $3x + 5y$ and $2x + 9y$
   b. $6x + 8y$ and $5x$
   c. $x + y + z$ and $3x + 9y + 15z$
2. Add the following $xy^2+5x^2y-8x^2y-3xy^2$ & xy^2+9x^2y
3. Add the following $3x^2y^2+9x^3y^3+14xy+14y$ & $12x^2y^2+19x^3y^3+12xy+12y$
4. Add these expressions $5xy+4x+8y$; $9xy+2x+18y$, $12xy+22x+17y$
5. Add $a^4-2a^3b^3+3a^2b^2+6ab$ and $-6a^4+8a^3b^3+5a^2b^2-12ab$
6. Subtract
   a. $4x$ from $3x$
   b. $2x^2$ from $-8x^2$
   c. $9xy$ from $12xy$
7. Subtract $12x^3-7x^2+6x-3$ from $22x^3+8x^2-15x+8$
8. Subtract $14a^4-21b^4+16ab+12a+ab$ from $36a^4+12b^4-14ab+15a+12b$
9. What should be subtracted from $2x^3+9x^2-8x+12$ to get $5x^3-8x^29x-18$

10. Add $9x^3+3x^2-6x$ to the sum of $5x^3+12x^2-13x+6$ and $13x^3+14x^2-11x+9$

Level 2: Exercise 5.7 Practice Questions (Addition and Subtraction)

Simplify

1. $2x+(6x-8y)$
2. $-9x-(3x-12)$
3. $-12xy+14xy(x-y)$
4. Simplify: 6a – (3b – 2c + a) + 2a (4b + 9ac + b)
5. Simplify: 2a - [3b {5a - (2b – 4c)}]
6. Simplify: -a - [b - {a - {(b - c) + 6a}]
7. Simplify: $30-[6xy+6\{x^2-(xy-y)-(x-y)\}]$
8. Simplify: $x^3-[3x^2+\{2x-(x^2-1)\}+6]$
9. Simplify: $9+[x-\{2y-(12x+y-4)+12x^2\}-(2x^2-4y)]$
10. Simplify: The sides of a rectangle are $3x^2+3x+7$ and $9x^2+5$. *Find its perimeter*
11. Simply: p – [p + {p + q – 2p - (m – 2q) – n]
12. Simplify: The sum of 2a – 3b and 4a – 6b + 9 is subtracted from 3a + 15b – 9

Level 3: Exercise 5.8 Practice Questions (Addition and Subtraction)

1. The perimeter of a scalene triangle is 30a + 9b – 6c. Two of its sides are 12a + 5b – 2c and -13a + 7b – 4c. Find the third side.
2. Add the terms $9x^8+6x^7-4x^6+3x^5-12x^4+6x^3+9x^2+5x-6$ and $-12x^8-6x^7+14x^6-13x^5-120x^4-16x^3-19x^2-15x-16$ and subtract from $20x^8+60x^7-14-3x^5+14x^4-16x^3+90x^2+100x-12$
3. If $A=-6x^2+9y^2+12xy$, $B=21x^2+18xy-9y^2$, $C=24x^2-26y^2+18xy$ Then find
   a. A + B + C
   b. A + C + C
   c. B + B + C
4. If $A=112x^2+13y^2+19xy$, $B=116x^2-4y^2-7xy$ and $C=122x^2-23y^2-17xy$ Find
   a. C – A + B

b. B - A + C
c. A - B - C

5. If $A = 16x^2 - 15x - 3xy$, $B = 12x^2 - 18x + 8xy$ and $C = 15x^2 - 36x + 9xy$, Find
a. 2A + 2B + 2C
b. A - 2B
c. A - 3C

5.5 Simple Linear Equation

5.5.1 Formation of Algebraic Equation

To 'compare two algebraic expressions' or 'find the variable's value in the algebraic expression,' we must use the equality operator "=" between two algebraic expressions whose values are equal.

For example, the age of Rahul's father is two times that of his son. Father's age after two years will be 45. Find the present age of Rahul.

The unknown factor Rahul's age is taken to be x; by the given statement two times Rahul's age + 2 years $(2x + 2)$ will be the father's age after two years = 45

Hence, $2x + 2 = 45$ is the Algebraic equation of the above situation, which has value on the Right-Hand side.

In the above example, instead of value, if the father's age is 20 years more than the son's age at present, the equation can be framed as $2x = x + 20$

Definition—Algebraic Equation

An 'Algebraic Equation' is a mathematical statement with two algebraic expressions on both sides of 'equal to' whose values are equal. Here, L.H.S. represents the algebraic expression on the Left-Hand side of the equation & R.H.S. represents the algebraic expression on the Right-Hand side of the equal to.

For Example, $3x + 4 = 1$, $6x^2 + 3 = 8x + 6$

The generalized form of Algebraic Equation

$ax + b = cx + d$

Algebraic Equation on L.H.S. Equal to Algebraic Equation on R.H.S.

For example: In $x^2 + 9x + 8 = (x + 1)(x + 8)$, $x^2 + 9x + 8$ is the L.H.S. Algebraic expression of the equation, $(x + 1)(x + 8)$ is the R.H.S. algebraic expression of the equation whose values are equal.

For example: A number that is multiplied by itself gives 9, which is converted to an algebraic equation as $x^2 = 9$.

For example: When we subtract 3 from a number, the result will be 8, which can be to the algebraic equation as $x - 3 = 8$

Definition—Degree of an equation

The highest power of the variable is the degree of an equation.

For Example: $3x + 4 = 1$, $6x^2 + 3 = 8x + 6$

Note: As discussed in the algebraic expression definition, the variables' exponent is a positive integer. Depending on the highest power of the variables referred to as the degree of the equation, they are named as follows.

LINEAR EQUATION—A Polynomial of degree one.
For Example: $2x + 5 = 10$

QUADRATIC EQUATION—A Polynomial of degree two.
For Example: $2x^2 + 3x + 5 = 0$

CUBIC EQUATION—A Polynomial of degree three.
For Example: $2x^3 + 3x^2 + x + 4 = 0$

Note: The degree of the non-zero constant polynomial is zero.

5.5.2 How to Solve Linear Algebraic Equations?

Suppose we want to find the variable's value in the algebraic equation. We have to find all the variables on the L.H.S. of the equation and find which values of the variable the value of L.H.S. is equal to the value of R.H.S. The process is termed solving the algebraic equation.

Definition—Transposition

Transposition—The process of isolating the required unknown-like terms to one side, especially on the L.H.S. of the equation, and the remaining known like-terms to the R.H.S. of the equation.

Example: $4x + 2 = 2x + 6$ or $4x - 2x = 6 - 2$

There are some rules we have to follow in rearranging the variables referred to as '**transposition**' from left to right or from right to left of the equal-to ("=") sign without affecting the equality of the equation.

Rules for transposition

Rule 1: We can add, subtract, multiply, or divide the same quantity into both sides of the equation.
Rule 2: Invert the expression on both sides of the equation.

For Example: $3x + 4 = 2x + 6$

$3x - 2x + 4 - 4 = 2x - 2x + 6 - 4$

$x = 2$, Here, subtract $2x$ & 4 on both sides of the equation to solve the equation.

For Example: $3x = 2x$ that is $\frac{1}{3x} = \frac{1}{2x}$

EXAMPLE 4: Solve $2x - 5 = x + 3$
SOLUTION: By transposition rules, we can subtract & subtract the same values on both sides of the equation,

$2x - x - 5 + 5 = x - x + 3 + 5 \rightarrow x = 8$

EXAMPLE 5: Find the number whose sum of 18 gives two times the number.
SOLUTION: Suppose the number is x. The algebraic equation from the given statement is $x + 18 = 2x$

By transposition rules, $x - x + 18 = 2x - x$

$$18 = x$$

On the other hand, instead of using transposition rules, we can rearrange the terms as per our requirement by changing the terms to inverse terms as per the operator.

Inverse operations in solving algebraic equations
Switching the operator before the term to the inverse operator can move the terms on one side of the equation to the other. The operator and their inverse operators are listed below.

Addition (+) ←→Subtraction (–)
Multiplication (x) ←→ Division (÷)
Power (^2) ←→ square root (√)

For Example: $4x + 3 = 2x + 6 \leftarrow 4x - 2x = 6 - 3$, here $+ 2x$ is shifted as $-2x$ from right to left & 3 is shifted as -3 from left to right.

EXAMPLE 6: The father and his son are currently 48 years old combined. The father's age is also two times that of his son. What are the father's and his son's ages?
SOLUTION: Suppose x is the son's age, so $2x$ is the father's age.

According to the given statement,

$$2x + x = 48$$

$$3x = 48$$

$$x = \frac{48}{3} = 16$$

Hence the father's age is 32 years and Son's age is 16 years.

Level 1: Exercise 5.9 Practice Questions (Linear Algebraic Equations)

Solve for x:

1. $3x+1=10$
2. $x+100=120$
3. $9x-12=-8$
4. $3(x-1)=24$
5. $3(x+1)=4(x-3)$
6. $\frac{x}{2}+\frac{1}{2}=\frac{3}{4}$
7. $2(x+1)-40=4(x-7)$
8. $\frac{9x}{3}-\frac{1}{3}=\frac{2}{3}$
9. $5(x-1)=\frac{1}{2}(x-3)$

Level 2: Exercise 5.10 Practice Questions (Linear Algebraic Equations)

1. $\frac{x}{2}+\frac{x}{3}=12$
2. $\frac{7}{x}+20=\frac{9}{12}$
3. $\frac{x}{2}-\frac{x}{5}+\frac{x}{10}=15$
4. $\frac{3}{7}(x-5)-\frac{4}{7}(x+6)=19$
5. $(x+2)(x-3)-(x+2)(x-6)=2(x-6)$
6. $\frac{y-1}{2}+\frac{y-2}{3}+\frac{y-3}{6}=18$

7. $m - \frac{(m-8)}{3} = -m + \frac{(m-9)}{27}$

8. $\frac{1}{3}\left(\frac{m-7}{7}\right) + \frac{1}{2}\left(\frac{m-8}{14}\right) = \frac{m-3}{8}$

9. $\frac{m-6}{7} + 2(m+6) = \frac{m-5}{8} + 3(m-6)$

10. $0.15(p-8) + 0.8(p-9) = 2.25$

Level 3: Exercise 5.11 Practice Questions (Linear Algebraic Equations)

Solve for x:

1. $[(3x+3)+(9x-5)]^2 = [(5x-3)+(7x-3)]^2$
2. $\frac{45-3x}{2} - \frac{35-2x}{6} = \frac{15-2x}{12} + \frac{18-9x}{24}$
3. $5\left(\frac{7x-18}{15}\right) - 8\left(\frac{9x+27}{12}\right) = 32\left(\frac{x+8}{12}\right)$
4. $6.8x + \left(\frac{6.8-2x}{9.2}\right) = 7.4 + \left(\frac{7.4-3x}{9.2}\right)$
5. $\frac{7x}{8} - \frac{x-2}{18} + 3\left(\frac{x-4}{4}\right) = \frac{12}{13}$

Level 1: Exercise 5.12 Practice Questions (Word Problems on Simple Linear Equations)

1. Six is added to the number that gives nine. Find the number.
2. If a number is multiplied by five and three is subtracted, the result is sixty. Locate the number.
3. Find the number that, when divided by 6, gives 360.
4. The total of the two numbers is twelve. One number is two more than another. Find the numerals.
5. What two numbers must be greater than one another by 9 and have a sum of 20?
6. Find the number that, when multiplied by 9, increases by 48.
7. The father and son are a total of 81 years old. Given that the father is twice as old as the son, how can you determine their ages?

8. John's father is three times John's age. He will be double John's age after 12 years. How old are they currently?
9. Abdul has a few chocolates. He gives half of them to his cousins and half to his friends. Now he has left with 6 chocolates. How many chocolates did Abdul have at first?
10. The sum of 2 consecutive numbers is 73. Find the numbers?

Level 2: Exercise 5.13 Practice Questions (Word Problems on Simple Linear Equations)

1. Neha's mother is 3 times as old as Neha and four times as old as Neha's sister Juhi. Juhi is 3 years younger than Neha. How old are Neha, Juhi, and their mother?
2. Irfan owns a rectangular-shaped block of land. He caged it in with 280 meters of wire. The plot is 5 m longer than it is wider. How far and how wide is the plot?
3. A purse only contains 25 Paise worth of currency. The amount of cash in the wallet is Rs. 12.50. How many 25 Paise coins are there?
4. Sam has twice the number of marbles as Tom. Tom has half the number of marbles as John. There are 210 marbles in all. Find the number of marbles Sam, Tom, and John have.
5. Arman is 2 times as old as his son. After 20 years, Arman will be 3/2 times as old as his son. Find their present ages.
6. Mrs. Janet pays her daughter's school fees which are thrice the fees paid for the extra classes. If the total fees paid per month is $150, then determine the monthly school fees.
7. During a discount sale, a shopkeeper sells one free shirt for every 3 shirts bought by the customers. Mr. Irfan gets 12 shirts in total. For how many shirts did he pay in total?
8. A college distributes prizes for the students. The second prize is $2/5^{th}$ of the first, and the third is 1/5th of the first. If the total value of the 3 prizes is $160, find the value of each award.
9. A contractor engages a worker for 30 days of work. He pays $20 per working day and deducts $3 for the worker's absence. The worker receives $462; how many days did he receive an absence?
10. Mrs. Julie invites her friends to her birthday party. Half the guests preferred drinking juice, and one-third chose soft

drinks. 25 invitees did not select either juice or soft drinks. How many friends did Julie invite?

Level 3: Exercise 5.14 Practice Questions (Word Problems on Simple Linear Equations)

1. 20 kg of an alloy of lead and tin contains 60% of lead. How much lead must be melted to make an alloy containing 75% lead?
2. The price of a table is twice that of a chair. The price of a cupboard is thrice the table. Mr. Jaffer pays $1400 to buy all three pieces of furniture. Find the cost of each type of furniture.
3. Mr. Abdul has $12000. He invests 10% in a deposit scheme, 20% in bonds, and the rest amount in stocks. Find the amount Abdul invested in stocks.
4. A.B.C.'s stock price doubles every 2 years. After 6 years, the stock would trade at $950. What was the stock price 6 years ago?
5. Omar attempts a mathematics exam. The exam had 100 multiple-choice questions. Each correct answer fetches 3 marks, and 1 mark is deducted for wrong or unattempted questions. Omar scored 220 marks on the exam. Determine how many questions Omar attempted correctly?

5.6 Quadratic Equations

The term "quadratic equation" refers to an algebraic equation with a degree of two.

The general form of the quadratic equation is as follows: $ax^2 + bx + c = 0$
Here, a is the coefficient of x^2, b is the coefficient of x and c is a constant
Example $3x^2 + 2x + 5 = 0$ is a quadratic equation.

5.6.1 Solving Quadratic Equation

There are two types of solving quadratic equations.

1) Factorization Method

(i) Bring all the terms to the equation's left side, leaving zero on the left side.

(ii) factorize L.H.S. by splitting the term with the variable x
(iii) Equate each factor with zero to find the value of x

EXAMPLE 7: Solve for x for the equation $x^2 + 4 = 4x$

SOLUTION: $x^2 - 4x + 4 = 0$

$$x^2 - 2x - 2x + 4 = 0$$
$$x(x-2) - 2(x-2) = 0$$
$$(x-2)(x-2) = 0$$
$$(x-2) = 0 \; or \; (x-2) = 0$$
$$x = 2,2$$

2) Method of Completing the Squares

$$(ax^2 + bx + c = 0)$$

1. Shift all the variable terms to the L.H.S. and constant to R.H.S.
2. Divide all the terms by the coefficient of x^2 to make it 1.
3. To form a perfect square on the L.H.S., add $\left(\frac{b^2}{2}\right)$ *on* both sides of the equation
4. By finding the square root on both sides, we can solve for x

EXAMPLE 8: Solve: $2x^2 - 5x + 3 = 0$
SOLUTION:

Step 1: $2x^2 - 5x = -3$ … [Shifting variable terms to L.H.S.]
Step 2: $x^2 - \frac{5}{2}x = -\frac{3}{2}$ … [Dividing all terms by 2]
Step 3: $x^2 - \frac{5}{2}x + \left(\frac{5}{4}\right)^2 = -\frac{3}{2} + \left(\frac{5}{4}\right)^2$ … [Add $\left(\frac{5}{4}\right)^2$ on both sides]

$$\left(x - \frac{5}{4}\right)^2 = -\frac{3}{2} + \frac{25}{16} = \frac{-24+25}{16} = \frac{1}{16}$$

Step 4: $x - \frac{5}{4} = (\pm)\frac{1}{4}$ …[Taking the square root of *both sides*]

$$x = \frac{5}{4} + \frac{1}{4} = \frac{6}{4} = \frac{3}{2} \quad \& \quad x = \frac{5}{4} - \frac{1}{4} = \frac{4}{4} = 1$$

$$x = 1, x = \frac{3}{2}$$

3) Quadratic Formula Method

By following the completing square method in the general quadratic equation

$$\mathrm{a}x^2 + \mathrm{b}x + \mathrm{c} = 0$$

We arrive with a formula.

$$x = \frac{-\mathrm{b} \pm \sqrt{\mathrm{b}^2 - 4\mathrm{ac}}}{2\mathrm{a}}$$

By analyzing the above formula, we have two distinct roots for x because of the $(\pm)$ signs. Also, the value inside the square root $(b^2 - 4ac)$ helps us to decide whether the nature of roots is real or imaginary.

Hence $\boldsymbol{b^2 - 4ac}$ is named as a discriminant.

Finding the nature of the roots:

In $ax^2 + bx + c = 0$

If the discriminant is

1) $b^2 - 4ac = 0$, then roots are real and equal
2) $b^2 - 4ac > 0$, then roots are distinct and real
3) $b^2 - 4ac < 0$, then roots are distinct and imaginary

EXAMPLE 9: Find the type of the equation's roots. $5x^2 + 2x + 8 = 0$

SOLUTION: Discriminant is $b^2 - 4ac$

$5x^2 + 2x + 8 = 0$

$\mathrm{a} = 5, \mathrm{b} = 2, \mathrm{c} = 8$

$b^2 - 4ac = 2^2 - 4(5)(8) = 4 - 160 = -156$

Hence $(b^2 - 4ac) < 0$ and the roots are imaginary and distinct.

EXAMPLE 10: Find the nature of roots $x^2 - 10x + 25 = 0$ and solve for x

SOLUTION: $x^2 - 10x + 25 = 0$, Here $a = 1$, $b = (-10)$, $c = 25$

$$b^2 - 4ac = (-10)^2 - 4(1)(25) = 100 - 100 = 0$$

Hence roots are equal and real.

$$\text{Also, } x = \frac{-b \pm \sqrt{b^2 - 4ac}}{2a}$$

$$= \frac{-b}{2a} \ldots \text{ (since equal roots)}$$

$$x = \frac{-(-10)}{2(1)} = 5$$

$x = 5, 5$ are the roots of the equation

EXAMPLE 11: The product of 2 consecutive numbers is 156. Find the numbers.
SOLUTION: Let x and $x+1$ be two consecutive the two consecutive numbers by given

$$(x)(x+1) = 156$$

$$x^2 + x = 156$$

$$x^2 + x - 156 = 0$$

$$x^2 + 13x - 12x - 156 = 0 \text{ (Factorization Method)}$$

$$x^2 + 13x - 12x - 156 = 0$$

$$x(x+13) - 12(x+13) = 0$$

$$\rightarrow (x+13)(x-12) = 0$$

$$\rightarrow x = -13, x = 12$$

The numbers are = -13, 12

EXAMPLE 12: Find the two numbers when the sum of a number and its reciprocal is $\frac{37}{6}$

SOLUTION: Let x be the number & and its reciprocal $\frac{1}{x}$

By given $x+\frac{1}{x}=\frac{37}{6}$

$\rightarrow \frac{x^2+1}{x}=\frac{37}{6}$

Cross multiply

$$6(x^2+1)=37x$$
$$6x^2-37x+6=0$$
$$6x^2-36x-x+6=0$$

By factorization method

$$6x(x-6)-(x-6)=0$$
$$(6x-1)-(x-6)=0$$
$$(6x-1)=0 \;\&\; (x-6)=0$$
$$x=\frac{1}{6}, x=6$$

The numbers are 6, $\frac{1}{6}$

Level 1: Exercise 5.15 Practice Questions (Quadratic Equation)

1. Determine the nature of the roots of the following equations
   a. $3x^2+x-1=0$
   b. $x^2+7x+2=0$
   c. $9x^2-x-3=0$

2. Find the value of k for which roots are real and equal.
   a. $3x^2 - 10x + k = 0$ for real and equal roots
   b. $9x^2 + 3kx + 6 = 0$
   c. $kx^2 - 6x + k = 0$
3. When a quadratic equation's x^2+mx+6=0 root is 3, find the value of k if two quadratic equations x^2+mx+k=0 have equal roots.
4. Find the value of k for which the equation $(k-3)x^2 + (k+1)(x+1) = 0$ has equal roots.
5. Determine the positive value of k for which the equation $x^2 + kx + 6 = 0$ has real and distinct roots.
6. Find out the least positive value of k for which the equation $x^2 + kx + 9 = 0$ has equal roots.\
7. Determine the nature of roots for the equation $(x-2m)(x-2n) = 2mn$
8. Solve the quadratic equations.
   a. $m^2 - 7m + 12 = 0$
   b. $3x^2 + 8x + 14 = 0$
   c. $24x(x+1) = 12$
9. Solve $\frac{x-6}{x-2} + \frac{x-7}{x-5} = \frac{11}{10}$
10. Solve $\frac{1}{x-6} - \frac{1}{x} = \frac{2}{5}$
11. The area of a square is 10.24 cm^2 determine the perimeter of the square.
12. Divide 30 into two parts such that one part is double the other.
13. Find the roots of the quadratic equation $6x^2 - 3\sqrt{6}\,x - 3 = 0$
14. If the sum of n natural numbers is 465, then find the value of n
15. The product of two positive integers is 600. Find the numbers
16. The dimensions of the rectangle are $(3x+6)$ and $(2x+5)$ units find the area of the rectangle for $x = 10\ units$.
17. Find the sides of a square whose area is given by $25x^2 + 125x + 25 = 0$ for $x = 3$
18. Find the two numbers whose sum is 42 and product 440.
19. Find 2 consecutive numbers whose sum of squares is 481.

Level 2: Exercise 5.16 Practice Questions (Quadratic Equation)

Solve the following Quadratic Equations

1. $x^2-(\sqrt{6}+1)x+\sqrt{3}=0$
2. $3\sqrt{5}m^2+15m-10\sqrt{5}=0$
3. $x^2-2\sqrt{7}x+7=0$
4. $8x^2-2\sqrt{16}x+2=0$
5. $\frac{m}{x-n}+\frac{n}{x-m}=3$
6. Solve using the quadratic formula $a^2x^2-6abx+3b^2=0$
7. $p^2x^2+(p^2-q^2)x-q^2=0p\neq 0$
8. Find the value of *of* $x^2-6x+k=0$ has distinct and real roots
9. Solve $\frac{1}{x+1}+\frac{3}{x+2}=\frac{7}{12}$, $(x\neq(1,-1)$
10. If - 6 is a root of the quadratic equation $3x^2+px-18=0$ and the quadratic equation $p(x^2+x)+k=0$ has equal roots to find the value of k.
11. The sum of reciprocals of David's age 8 years ago and 8 years from now is $\frac{11}{24}$. Find the present age.
12. The sum of natural numbers is 5 and the sum of their reciprocals is $\frac{5}{6}$. Find the numbers.
13. Determine the positive value of k for which $x^2+4x+64=0$ and $x^2-8x+k=0$ will both have real roots.
14. The product of 2 consecutive even numbers is 440 find the numbers.
15. The difference in the square of the two numbers is 700. The larger number is 10 more than the smaller number.
16. The sum of the areas of 2 squares is 369 m^2, if the difference of their perimeters is 12 m. Find the sides of the two squares.
17. Find the natural differences by 2 and the sum of whose square is 100.
18. Joe and John together have 30 marbles. Both of them gave 6 marbles to each of their friend Jack. The product of the number of marbles they have is 65. Find the number of marbles they had first.

19. A two-digit number is such that the product of digits is 8, When 18 is subtracted from the marbles, two digits get interchanged to find the number.

Level 3: Exercise 5.17 Practice Questions (Quadratic Equation)

1. Mr. Jackson sells a toy for Rs $75 and as much as a percent of the cost price of the toy. Find the cost price of the toy.
2. A herd of cows was seen to be two-fifths in the woodland. The cow herd's square root had moved to the riverfront. The final ten cows were close to the mountain. Determine the number of cows overall in the woodland.
3. Alisha plans a tour for $600. If she extends her tour for 10 days, she must cut her daily expense by $10. Find the original duration of the tour.
4. An isosceles triangle has a surface area of 24 cm^2 with equal sides 10 cm long. Identify the foundation of it.
5. If Rubina were younger by 5 years than she is, then the square of her age (in years) would have been 11 years more than 5 times her actual age. What is her age now?
6. For $400, Mr. Jack takes a tour. He must reduce daily spending by $2 if he extends his tour for 30 days. Where can he find the time?
7. For the same amount of work, B takes two hours more than A to complete the work; find how much time B will take if both A and B can complete the work in 12 hours.
8. For what value of k does the quadratic equation below have equal roots?

5.7 Simultaneous Equations

In some situations of converting the word problems to algebraic equations, we have arrived at two or more equations regarding the unknown variable; these equations are termed simultaneous equations (or) systems of simultaneous equations. We have to notice that for 'n' unknown variables, there should be a minimum of 'n' simultaneous equations to find the variable's value. There are three methods to solve simultaneous equations.

(i) Elimination method by addition or subtraction.
(ii) Substitution method
(iii) Graphical method

The general form of a simultaneous linear equation in two variables is

$$a_1x + b_1y + c_1 = 0$$
$$a_2x + b_2y + c_2 = 0$$
$$.$$
$$.$$
$$.$$
$$.$$
$$a_nx + b_ny + c_n = 0$$
$$x, y \text{ are two variables}$$

$a_{1,}a_2 \ldots, a_n$ are coefficients of x
$b_1, b_2, \ldots b_n$ are coefficients of y
$c_1, c_2, \ldots c_n$ are constants.

i) Elimination Method by Addition or Subtraction:

1. Make the coefficient of any one variable the same in any two equations by multiplying the required number all over the equation.
2. Add or subtract two equations to eliminate the variable in L.H.S. & R.H.S.
3. Solve the remaining equation for another variable.
4. We have to repeat the process if there are three variables with three simultaneous equations by combining any two equations

EXAMPLE 13: Solve the given system of linear equations.

$$3x + 2y = 13 \qquad \text{...(1)}$$

$$5x - 3y = 9 \qquad ...(2)$$

SOLUTION: To eliminate x, multiply equation (1) by 5 & equation (2) by 3

$$\begin{aligned} &\text{equation } (1)\times 5 && 15x + 10y = 65 \\ &\text{equation } (2)\times 2 && 15x - 6y = 27 \\ &\text{subtract } (1)\text{-}(2) && - \quad + \qquad - \end{aligned}$$

$$16y = 32$$

$$y = 2$$

Substitute y = 2 in equation (1)

$$3x + 2y = 13$$
$$3x + 2(2) = 13$$
$$3x = 13 - 4$$
$$3x = 9$$
$$x = 3$$

Hence $x = 3$, $y = 2$ are two solutions of simultaneous linear equation in two variables

(ii) Substitution Method:
In some situations, equations are easily solved by substituting one equation with other

EXAMPLE 14: Solve for x and y.

$$x = y + 6 \qquad ...(1)$$

$$2x + 3y = 27 \qquad ...(2)$$

SOLUTION: Substitute (1) in equation (2)

$$2(y+6)+3y=27$$
$$2y+12+3y=27$$
$$5y+12=27$$
$$5y=27-12$$
$$5y=15$$
$$y=\frac{15}{5}$$
$$y=3$$

Substitute $y=3$ in equation (1)

$$x=3+6$$
$$x=9$$

x = 9, y = 3 are the solutions of the equation.

EXAMPLE 15: Solve the following system of linear equations by using the method of elimination

$$3x+2y=10 \qquad \ldots (1)$$

$$2\,x+3y=12 \qquad \ldots (2)$$

SOLUTION: To eliminate x we have to multiply

Equation $(1)\times 2$	$6x+4y=20$
Equation $(2)\times 3$	$6x+9y=36$
Subtract	$(-)\ (-)\ (-)$

$$-5y=-16$$
$$y=\frac{16}{5}$$

Substituting $y=\frac{16}{5}$ in equation (1), we get,

$$3x+2y=10$$
$$3x+(2)\frac{16}{5}=10$$
$$3x+\frac{32}{5}=10$$
$$15x+32=50$$
$$15x=50-32$$
$$15x=18$$
$$x=\frac{18}{15}=\frac{6}{5}$$

EXAMPLE 16: Solve the following system of linear equations by using the method of elimination.

$$9x+5y=18 \qquad \ldots (1)$$

$$15y=12x=24 \qquad \ldots (2)$$

SOLUTION: Here we can easily eliminate y by multiplying equation (1) by 3

$$27x+15y=54$$
$$12x+15y=24$$
$$\text{Subtracting } (-) \quad (-) \quad (-)$$

$$15x=30$$
$$x=\frac{30}{15}=2$$

Substitute $x = 2$ in $9x + 5y = 18$

$9(2) + 5y = 18$

$18 + 5y = 18$

$5y = 0$

$y = 0$

$x = 2, y = 0$

EXAMPLE 17: Solve

$$12x - 8y = 15 \quad \dots (1)$$

$$3x + 4y = 12 \quad \dots (2)$$

SOLUTION: We can eliminate x by multiplying equation (2) by 4

Equation (1)	$12x - 8y = 15$
Equation (2)×	$12x - 16y = 48$
subtracting	(–) (–) (–)

$$24y = -33$$

$$y = \frac{33}{24} = \frac{11}{8}$$

Substitute $y = \frac{11}{8}$ in equation $12x - 8y = 15$

$12x - 8\left(\frac{11}{8}\right) = 15$

$12x - (11) = 15$

$12x - (11) = 15$

$12x = 26$

$x = \frac{26}{12} = \frac{13}{6}$

$x = \frac{13}{6}$ and $y = \frac{11}{8}$ are two solutions

EXAMPLE 18: Solve for x and y

$$\frac{x}{10}+\frac{y}{5}=25 \qquad \ldots(1)$$

$$\frac{x}{20}-\frac{y}{4}=30 \qquad \ldots(2)$$

SOLUTION:

$$\frac{x}{10}+\frac{y}{5}=25$$
$$\frac{x+2y}{10}=25 \qquad \ldots(1)$$
$$x+2y=250$$

$$\frac{x}{20}-\frac{y}{4}=30$$
$$\frac{x-5y}{20}=30 \qquad \ldots(2)$$
$$x-5y=600$$

Equation $(1)-(2)$ $\quad x+2y=250$

$\quad x-5y=600$

subtracting $\quad (-) \quad (+) \quad (-)$

$7y=-350$

$$y=\frac{-350}{7}=-50$$

$$\text{substitute } y = -50 \text{ in } x + 2y = 250$$
$$x + 2(-50) = 250$$
$$x = 250 + 100$$
$$x = 350,\ y = -50$$

EXAMPLE 19: Solve for *x* and y.

$$3x - \frac{5}{2}y = 12 \qquad \ldots (1)$$

$$9x + \frac{8}{2}y = 15 \qquad \ldots (2)$$

SOLUTION:

$$6x - 5y = 24$$
$$18x + 8y = 30$$

To eliminate *x*, we have to multiply.

$$\begin{array}{rl} \text{Equation } (1) \text{ by } 3 & 18x - 15y = 72 \\ (2) & 18x + 8y = 30 \\ \text{subtracting} & (-)\ \ (-)\ \ (-) \end{array}$$

$$-23y = 42$$

$$y = \frac{-42}{23}$$

$$\text{substitute } y = \frac{-42}{23} \text{ in } 6x - 5y = 24$$

$$6x - 5\left(\frac{-42}{23}\right) = 24$$

$$138x + 210 = 55$$

$$138x = 552 - 210$$

$$138x = 342$$

$$x = \frac{342}{138} = \frac{171}{69} = \frac{57}{23}$$

$$x = \frac{57}{23}, y = \frac{-42}{23} \text{ are the solutions}$$

EXAMPLE 20: solve for *x* and y.

$$\frac{1}{2x} + \frac{1}{y} = \frac{1}{4} \qquad \ldots (1)$$

$$\frac{1}{3x} - \frac{1}{2y} = \frac{1}{8} \qquad \ldots (2)$$

SOLUTION:

$$\frac{1}{2x} + \frac{1}{y} = \frac{1}{4}$$

$$\frac{y + 2x}{2xy} = \frac{1}{4} \qquad \ldots (1)$$

$$4y + 8x = 2xy \ldots \text{(Cross Multiply)}$$

$$2y + 4x = xy \ldots \text{ divide by 2 } \ldots$$

$$\frac{1}{3x} - \frac{1}{2y} = \frac{1}{8}$$

$$\frac{2y - 3x}{6xy} = \frac{1}{8} \qquad \ldots (2)$$

$16y - 24x = 6xy$...(Cross Multiply)

$8y - 12x = 3xy$...(Divide by 2 into all terms)

Multiply by 3 in equation (1) $6y + 12x = 3xy$

(2) $8y - 12x = 3xy$

Adding $14y = 6xy$

$6x = 14$

$$x = \frac{7}{3}$$

Substitute $x = \frac{7}{3}$ in $2y + 4x = xy$

$$2y + 4\left(\frac{7}{3}\right) = \left(\frac{7}{3}\right)y$$

$6y + 28 = 7y$

$7y - 6y = 28$

$y = 28$

$x = \frac{7}{3}, y = 28$ are the solutions.

EXAMPLE 21: solve for *x* and y.

$$2x + \frac{5}{y} = 20 \qquad \ldots\ldots (1)$$

$$6x+\frac{8}{y}=24 \qquad \text{.....(2)}$$

SOLUTION:

$$\text{Equation} \quad (1)\times 3 \quad 6x+\frac{15}{y}=60$$

$$(2) \quad 6x+\frac{8}{y}=24$$

$$\text{subtracting} \quad (-) \quad (-) \quad (-)$$

$$\frac{15}{y}-\frac{8}{y}=36$$

$$36y=7$$

$$y=\frac{7}{36}$$

substituting $y=\frac{7}{36}$ in $2x+\frac{5}{y}=20$

$$2x+\frac{5}{\frac{7}{36}}=20$$

$$14x+180=140$$

$$14x=140-180=-40$$

$$14x=-40$$

$$x=\frac{-40}{14}$$

$$x=\frac{-20}{7}$$

$x=\frac{-20}{7}, y=\frac{7}{36}$ are the solutions.

EXAMPLE 22: solve for x and y.

$$\frac{5}{x+y}-\frac{2}{x-y}=-2 \qquad \ldots (1)$$

$$\frac{12}{x+y}+\frac{6}{x-y}=12 \qquad \ldots (2)$$

SOLUTION: Take $u=x+y, and\ v=x-y$

$$\frac{5}{u}-\frac{2}{v}=-2 \qquad \ldots (1)$$

$$\frac{12}{u}+\frac{6}{v}=12 \qquad \ldots (2)$$

Equation $(1)\times 3$ $\quad \frac{15}{u}-\frac{6}{v}=-6$

(2) $\quad \frac{12}{u}+\frac{6}{v}=12$

By adding ----------------

$$\frac{27}{u}=6$$

$$u=\frac{27}{6}=\frac{9}{2}$$

Substituting $u=\frac{9}{2}$ in $\frac{5}{u}-\frac{2}{v}=-2$

$$\frac{5}{\frac{9}{2}}-\frac{2}{v}=-2$$

$$\frac{10}{9}+2=\frac{2}{v}$$

$$\frac{28}{9} = \frac{2}{v}$$

$$v = \frac{2 \times 9}{28}$$

$$v = \frac{9}{14}$$

$$u = x + y,$$

$$x + y = \frac{9}{2} \qquad \ldots (3)$$

$$v = x - y$$

$$x - y = \frac{9}{14} \qquad \ldots (4)$$

Adding equations (3) and (4), we get,

$$x + y = \frac{9}{2}$$

$$x - y = \frac{9}{14}$$

$$2x = \frac{9}{2} + \frac{9}{14}$$

$$2x = \frac{63 + 9}{14}$$

$$2x = \frac{72}{14}$$

$$x = \frac{72}{28} = \frac{36}{14} = \frac{18}{7} = \frac{18}{7}$$

Substitute $x = \frac{18}{7}$ in $x + y = \frac{9}{2}$

$$\frac{18}{7} + y = \frac{9}{2}$$

$$y = \frac{9}{2} - \frac{18}{7}$$

$$y = \frac{63 - 36}{14}, \mathrm{y} = \frac{27}{14}$$

Hence $x = \frac{18}{7}, y = \frac{27}{14}$ are the solutions

EXAMPLE 23: Find the values of x, y and z.

$$x + 2y + z = 7 \qquad \ldots (1)$$

$$x + 3z = 12 \qquad \ldots (2)$$

$$2x - 3y = 2 \qquad \ldots (3)$$

SOLUTION:

Equation $(1) \times 3$ $\quad 3x + 6y + 3z = 21$

Equation (2) $\quad x + 0 + 3z = 12$

subtracting $\quad (-)\ (-)\ (-)\ (-)$

$$2x + 6y = 9$$

Equation (3) $\quad 2x - 3y = 2$

subtracting $\quad (-)\ (+)\ (-)$

$9y = 7 \quad y = \frac{7}{9}$

Substitute $y = \frac{7}{9}$ in $2x - 3y = 2$

$$2x - 3\left(\frac{7}{9}\right) = 2$$

$$2x = 2 + \frac{7}{3}$$

$$2x = \frac{13}{3}, x = \frac{13}{6}$$

Substitute x & y in equation (1)

$$x + 2y + z = 7$$

$$\frac{13}{6} + 2\left(\frac{7}{9}\right) + z = 7$$

$$z = 7 - \frac{13}{6} - \left(\frac{14}{9}\right)$$

$$z = \frac{2X63 - 39 - 28}{18}$$

$$z = \frac{59}{18}$$

$x = \frac{13}{6}, y = \frac{7}{9}, z = \frac{59}{18}$ are the solutions.

EXAMPLE 24: Find the value of u and v in the given equation.

$$\frac{15}{u} + \frac{12}{v} = 10 \qquad \ldots (1)$$

$$\frac{1}{u} + \frac{1}{v} = 12 \qquad \ldots (2)$$

SOLUTION:

$$\text{Equation}(1) \qquad \frac{15}{u}+\frac{12}{v}=10$$

$$\text{Equation}(2)\times 12 \qquad \frac{12}{u}+\frac{12}{v}=144$$

$$\text{subtracting} \qquad (-)\ (-)\ (-)$$

$$\frac{3}{u}=-134$$

$$u=\frac{-3}{134}$$

$$\text{Substitute } u=\frac{-3}{134} \text{ in } \frac{1}{u}+\frac{1}{v}=12$$

$$\frac{1}{\frac{-3}{134}}+\frac{1}{v}=12$$

$$\frac{1}{v}=12+\frac{134}{3}$$

$$\frac{1}{v}=\frac{36+134}{3}$$

$$\frac{1}{v}=\frac{170}{3}$$

$$v=\frac{3}{170}$$

$$u=\frac{-3}{134},\ v=\frac{3}{170} \text{ the solutions}$$

EXAMPLE 25: Solve

$$\frac{1}{3(2x+3y)}+\frac{12}{6(3x-2y)}=\frac{1}{2} \qquad \ldots(1)$$

$$\frac{6}{(2x+3y)}+\frac{4}{(3x-2y)}=2 \qquad \dots (2)$$

SOLUTION: Consider $u=2x+3y, v=3x-2y$

Therefore, we get the following equations.

$$\frac{1}{3(u)}+\frac{2}{(v)}=\frac{1}{2} \qquad \dots (1)$$

$$\frac{6}{(u)}+\frac{4}{(v)}=2 \qquad \dots (2)$$

Equation$(1)\times 2$ $\quad \frac{2}{3(u)}+\frac{4}{(v)}=1$

Equation(2) $\quad \frac{6}{(u)}+\frac{4}{(v)}=2$

subtracting $\quad (-) \quad (-) \quad (-)$

$$\frac{2}{3(u)}-\frac{6}{(u)}=-1$$

$$\frac{2-18}{3(u)}=-1$$

$$\frac{-16}{3(u)}=-1$$

$$u=\frac{16}{3}$$

Substitute $u=\frac{16}{3}$ in $\frac{6}{(u)}+\frac{4}{(v)}=2$

$$\frac{6}{\left(\frac{16}{3}\right)}+\frac{4}{(v)}=2$$

$$\frac{4}{(v)}=2-\frac{18}{(16)}$$

$$\frac{4}{(v)} = \frac{32-18}{(16)}$$

$$\frac{4}{(v)} = \frac{14}{(16)}$$

$$v = \frac{(16)\times 4}{14} = \frac{32}{7}$$

$$2x+3y = \frac{16}{3} \quad \ldots.. \quad \times 3$$

$$3x-2y = \frac{32}{7} \quad \ldots.. \quad \times 2$$

$$6x+9y=16 \qquad \ldots (3)$$

$$6x-4y = \frac{64}{7} \qquad \ldots (4)$$

subtracting (–) (+) (–)

$$13y = \frac{16\times 7-64}{7}$$

$$y = \frac{48}{91}$$

Substitute $y = \frac{48}{91}$ in $3x-2y = \frac{32}{7}$

$$3x-2\left(\frac{48}{91}\right) = \frac{32}{7}$$

$$3x = \frac{32}{7} + \left(\frac{96}{91}\right)$$

$$3x = \frac{32\times 3+96}{91} = \frac{416+96}{91}$$

$x = \frac{512}{273}$ *and* $y = \frac{48}{91}$ are the solutions.

EXAMPLE 26: solve.

$$3(6u+v)=14uv \quad \ldots (1)$$

$$3(9u+4v)=11uv \quad \ldots (2)$$

SOLUTION:

Equation $(1)\div uv$ $\quad \frac{18u}{uv}+\frac{3v}{uv}=14 \quad \frac{18}{v}+\frac{3}{u}=14 \quad \ldots (3)$

Equation $(2)\div uv$ $\quad \frac{27u}{uv}+\frac{12v}{uv}=11 \quad \frac{27}{v}+\frac{12}{u}=11 \quad \ldots (4)$

Equation $(3)\times 4$ $\quad \frac{72}{v}+\frac{12}{u}=56\ldots(5)$

Equation (2) $\quad \frac{27}{v}+\frac{12}{u}=11\ldots(6)$

subtracting $\quad (-) \quad (-) \quad (-)$

$$\frac{45}{v}=45$$

$v=1$

Substitute $\quad v=1$ in $\frac{27u}{v}+\frac{12}{u}=11$

$27+\frac{12}{u}=11$

$\frac{12}{u}=11-27=-16$

$u=\frac{-12}{16}=\frac{-3}{4}$

$u=\frac{-3}{4}$

$u=\frac{-3}{4}$, $v=1$ are the solutions

EXAMPLE 27: Suppose the solutions to the following linear equation system are infinite. Determine the value of k.

$$x+(k+1)y=5 \qquad \text{.......(1)}$$

$$(k+1)x+9y=8k-1 \qquad \text{.......(2)}$$

SOLUTION: The condition that linear equations have infinite solutions is that

$$\frac{a_1}{a_2}=\frac{b_1}{b_2}=\frac{c_1}{c_2}$$

Here $\frac{1}{(k+1)}=\frac{(k+1)}{9}=\frac{5}{8k-1}$

Comparing the first two terms,

$$9=(k+1)^2$$

$$k^2+2k+1-9=0$$

$$k^2+2k-8=0$$

$$k^2-2k+4k-8=0$$

$$k(k-2)+4(k-2)=0$$

$$(k+4)(k-2)=0$$

$$k=-4,\ k=2$$

Comparing the last two terms

$$\frac{(k+1)}{9}=\frac{5}{8k-1}$$

$$(k+1)(8k-1)=45$$

$$8k^2+8k-k-1=45$$

$$8k^2+7k-46=0$$

$$k = \frac{-7 \pm \sqrt{7^2 - (4)(8)(-46)}}{16}$$

$$k = \frac{-7 \pm \sqrt{49 + 1472}}{16}$$

$$k = \frac{-7 \pm \sqrt{1521}}{16}$$

$$k = \frac{-7 \pm 39}{16}$$

$$k = \frac{-7 - 39}{16}, k = \frac{-7 + 39}{16}, k = \frac{-46}{16}, k = \frac{32}{16}$$

The common answer k=2 is the value for which the linear equations have infinite solutions.

$$k = \frac{-46}{16} \quad k = 2$$

EXAMPLE 28: Suppose the following system of linear equations has no solutions. Determine the value of k.

$$(3k+1)x + 3y - 2 = 0 \qquad \text{.......} (1)$$

$$(k^2+1)x + (k-2)y - 5 = 0 \qquad \text{.......} (2)$$

SOLUTION: Condition for the system of the equation to have no solution is.

$$\frac{a_1}{a_2} = \frac{b_1}{b_2} \neq \frac{c_1}{c_2}$$

Hence, $\frac{3k+1}{k^2+1} = \frac{3}{k-2} \neq \frac{-2}{-5}$

Comparing the first two terms,

$$\frac{3k+1}{k^2+1}=\frac{3}{k-2}$$
$$(3k+1)(k-2)=3(k^2+1)$$
$$3k^2+k-6k-2=3k^2+3-5k=5$$

Clearly $\frac{3}{k-2}\neq\frac{-2}{-5}$ for $k=-1$

EXAMPLE 29: Solve

$$\frac{x}{a}+\frac{y}{b}=a+b \qquad \ldots(1)$$

$$\frac{x}{a^2}+\frac{y}{b^2}=3 \qquad \ldots\ldots..(2)$$

SOLUTION:

Multiply $\frac{1}{a}$ in equation (1) $\quad \frac{x}{a^2}+\frac{y}{ab}=\frac{a+b}{a}$

Equation (2) $\quad \frac{x}{a^2}+\frac{y}{b^2}=3$

subtracting $\quad (-)\ (-)\quad (-)$

$$y\left(\frac{1}{ab}-\frac{1}{b^2}\right)=\frac{a+b}{a}-3$$

$$y\left(\frac{b-a}{ab^2}\right)=\frac{a+b-3a}{a}$$

$$y=\frac{-2a+b}{a}\times\left(\frac{ab^2}{b-a}\right)$$

$$y=\left(\frac{b^2(b-2a)}{b-a}\right) \text{ where } b\neq a$$

Substitute in equation (1) $\frac{x}{a}+\frac{y}{b}=a+b$

$$\frac{x}{a}+\frac{b^2(b-2a)}{b(b-a)}=a+b$$

$$\frac{x}{a}=a+b-\frac{b(b-2a)}{(b-a)}$$

$$\frac{x}{a}=\frac{b^2-a^2-b^2+2ab}{(b-a)}$$

where $b \neq a,\ x=\dfrac{a^2(2b-a)}{b-a}$

$$x=\frac{a^2(2b-a)}{b-a}\ y=\left(\frac{b^2(b-2a)}{b-a}\right)$$

EXAMPLE 30: The two numbers' differences are 30, and their aggregate is 60. find the numbers.

SOLUTION: Let x and y be the two numbers

By given

$$x+y=60 \qquad \text{.... (1)}$$

$$x-y=30 \qquad \text{.... (2)}$$

Add (1) and (2)

$2x= 90$

$x = 45$

Sub, $x = 45$ in $x + y= 60$

$45 + y= 60$

$y= 60 - 45$

$y= 60 - 45$

$y = 15$

The numbers 45 and 15.

EXAMPLE 31: The sum of two numbers exceeds the smaller by 18. If the difference between them is 2, find the number.

SOLUTION: Let x and y be two numbers (x smaller and y larger)

By given,

$$x + y = x + 18 \quad \ldots (1)$$

y = 18
By given,

$$y - x = 2 \quad \ldots (2)$$

18 – x = 2
$x = 18 - 2$
$x = 16$
The two numbers are 16 and 18.

EXAMPLE 32: In a piggy bank, two types of coins are worth $6000. The number of coins worth $3 is 80 more than $5. Find the number of coins for each type.
SOLUTION: Let x and y be the numbers of coins worth $3 and $5, respectively
By given,

$$3x + 5y = 6000 \quad \ldots (1)$$

$$x = y + 80$$

$$x - y = 80 \quad \ldots (2)$$

$$\text{Equation } (2) \times 3 \rightarrow 3x - 3y = 240 \quad \ldots (3)$$

Subtract (3) from (1)
$3x + 5y = 6000$
$3x - 3y = 240$
(-) (-) (-)

8y= 5760
y = 720
Substitute y = 720 in $x - 720 = 80$
$x = 720 + 80$
$x = 800$
The number of coins worth $3 is 800
The number of coins worth $5 is 720

EXAMPLE 33: Two people have a 4:5 income-to-expenditure ratio and a 1:2 income-to-expenditure ratio. Determine their monthly incomes if they can save Rs. 9000 each month.

SOLUTION: Let $4x$ and $5x$ be the incomes of two persons, and y and 2y are the expenditures of two persons.

$$4x - y = 9000 \qquad \ldots (1)$$

$$5x - 2y = 9000 \qquad \ldots (2)$$

Equation (1) ×2

$(8x - 2y) = 18000$

$5x - 2y = 9000$

(-) (+) (-)

$3x = 9000$

$x = 3000$

Substitute $x = 3000$ in $4x - y = 9000$

$4(3000) - y = 9000$

$12000 - y = 9000$

$y = 3000$

Their monthly incomes are $4x = 12000$ and $5x = 15000$, respectively.

EXAMPLE 34: 55 is obtained by adding a two-digit number with the outcome of reversing the number. Instead, if the digits differ by 1, find the number.

SOLUTION: Let x, and y be a two-digit number.

Therefore, $10x+y$ is the expanded form. Reversing the digits, y x is the new number.

By given,

$$10x + y + 10y + x = 55$$

$$11x + 11y = 55$$

$$x + y = \frac{55}{11} \qquad \ldots (1)$$

$$x + y = 5$$

Also, by giving,

$$y - x = 1 \qquad \dots (2)$$

Adding equations (1) and (2), we get,

$y - x = 1$

$y + x = 5$

$2y = 6$

$y = 3$

Substituting $y=3$ in $y - x = 1$

$3 - x = 1$

$x = 3 - 1$

$x = 2$

Hence the required digits are $x = 2$ and $y = 3$, and the number is 23.

EXAMPLE 35: Five years ago, a man was 4 times as old as his son, while five years later, the difference between their ages is 30. Find their present ages.

SOLUTION: Five years ago, the age of man was $4x$, and his son was x

Hence, the present age of man is $4x+5$

Hence, the present age of his son is $x+5$

Five years later, their ages are $4x+10$ and $x+10$, respectively. By given,

$(4x+10) - (x+10) = 30$

$4x + 10 - x - 10 = 30$

$3x = 30$

$x = 10$

Present age of the man $= 4x+5 = 45$

The present age of his son is $x+5 = 15$

EXAMPLE 36: A two-person household's income-to-expense ratio is 9:7 and 4:3, respectively. Suppose each can save $2,000 every month. Find what they make each month.

SOLUTION: Let $9x$ and $7x$ be the incomes of the persons Let the ratio of expenditure be 4y and 3y for the two persons. By given,

$9x - 4y = 2000$ ... (1) ... (since income-expenditure = savings)

$7x - 3y = 2000$... (2)

$$\text{Equation } (1)\times 3 \quad \rightarrow 27x-12y=6000$$
$$\text{Equation } (2)\times 4 \quad \rightarrow 28x-12y=8000$$
$$(-)\quad (+)\qquad (-)$$

$$-x=-2000$$

$$x=2000$$

Substituting $x = 2000$ in incomes $9x$ and $7x$

$$9\times 2000=18000 \text{ and}$$
$$7\times 2000=14000 \text{ are their incomes}$$

EXAMPLE 37: Find the area of the rectangle with a perimeter of 28 feet and length and breadth that are 4:3.
SOLUTION: Let l and b be the length and breadth of the rectangle
By given, $2(l+b) = 28$

$$2l+2b=28 \qquad \dots (1)$$

Also, $l = 4x$ and $b = 3x$ respectively
Substituting in (1) we get,

$$2l+2b=2\times 4x+2\times 3x=28$$
$$8x+6x=28$$
$$14x=28$$
$$x=2$$

Hence length $4x=8$ and breadth $3x=3\times 2=6$

Area $=6\times 8=48$ sq. feet

EXAMPLE 38: One pizza and a soda cost \$6 in a restaurant. 3 pizzas and 2 sodas will set you back \$16.50. Find out how much a pizza and a soda are worth.
SOLUTION: Let x and y be the cost of a Pizza and a Coke, respectively.
By given

$$x + y = 6 \qquad \ldots (1)$$

$$3x + 2y = 16.5 \qquad \ldots (2)$$

$$(1)\times 3 \qquad 3x + 3y = 18 \quad \ldots\ldots(3)$$

$$3x + 2y = 16.5 \quad \ldots(2)$$

$$-y = -1.5 \text{ (by subtraction)}$$

$$y = 1.5$$

Substituting y = 1.5 in equation (1)

$$x + 1.5 = 6$$

$$x = 6 - 1.5$$

$$x = 4.5$$

The cost of pizza is \$4.5 & coke is \$1.5

EXAMPLE 39: Five years hence, age of Sam will be three times that of his son. Five years ago, Sam's age was seven times that of his son. What are their percentages?

SOLUTION: Let x and y be the present age of Sam and his son. Five years hence x +5 and y+5 are their ages.

$$\text{Sam's age } (x+5) = 3(y+5)$$

$$x - 3y = 10 \qquad \ldots\ldots\ldots (1)$$

Five years ago, (x – 5) and (y – 5) were their ages.
By given,

$$(x-5) = 7(y-5)$$

$$x - 7y = -30 \qquad \ldots\ldots\ldots (2)$$

Subtracting equation (1) from equation (2), we get,

$$x-7y=-30$$
$$x-3y=10$$
$$(-)\quad(+)\qquad(-)$$

$$-4y=-40$$
$$y=10$$
$$\text{Substitute } y=10 \text{ in } x-3y=10$$
$$x-30=10$$
$$x=40$$

At present, Sam's age is 40, and his son's age is 10 years

Mixed Level: Exercise 5.18 Practice Questions (Simultaneous Equations)

Solve the following system of equations.

1. $\sqrt{2}x-\sqrt{3}y=0$
   $\sqrt{3}x-\sqrt{7}y=0$
2. $0.3x+0.2y=0.5$
   $0.2x+0.3y=0.75$
3. $\dfrac{22}{x+y}+\dfrac{15}{x-y}=6$
   $\dfrac{55}{x+y}-\dfrac{12}{x-y}=9$
4. $\dfrac{3}{x}+\dfrac{2}{3y}=\dfrac{1}{9}$
   $\dfrac{2}{x}+\dfrac{3}{y}=0$
5. $5x+\dfrac{9}{y}=18$
   $9x-\dfrac{8}{y}=24$

6. $6v - 3u = 9uv)$

   $8v - 5u = 12uv$
7. $7(y+2) - 2(x-2) = 12$

   $4(y-2) + 3(x-1) = 2$
8. Find the value of k for which the initial system of equations has infinitely many solutions.
   a. $kx + 3y - k = 3$

      $12x + ky - k = 0$
   b. $6x + 2y = k$

      $x + 4y = 3$
9. Find the value of k for which the system has no solution.

$$2x - ky + 3 = 0$$
$$6x - 4y - 1 = 0$$

10. Find the value of a and b for which the following system of the linear equation has infinite solutions

$$3x - (a-3)y = 2b + 1$$
$$4x - (a-1)y = 5b - 1$$

5.8 Graphs of Algebraic Functions

5.8.1 Graph of Linear Equations

Linear functions are those whose graph is a straight line. A linear function has the following form. y = f(x) = a + bx. A linear function has one independent variable and one dependent variable.

Here x is called the independent variable while y is the dependent variable.

EXAMPLE 40: Draw the graphs of the following:

a) x=0
b) y =0
c) x =2
d) y=3

e) $2x - 3 = 12$
f) $\frac{1}{2}x - 3 = 3$
g) $x - y = 6$

SOLUTION:

(**a**) The graph for x = 0

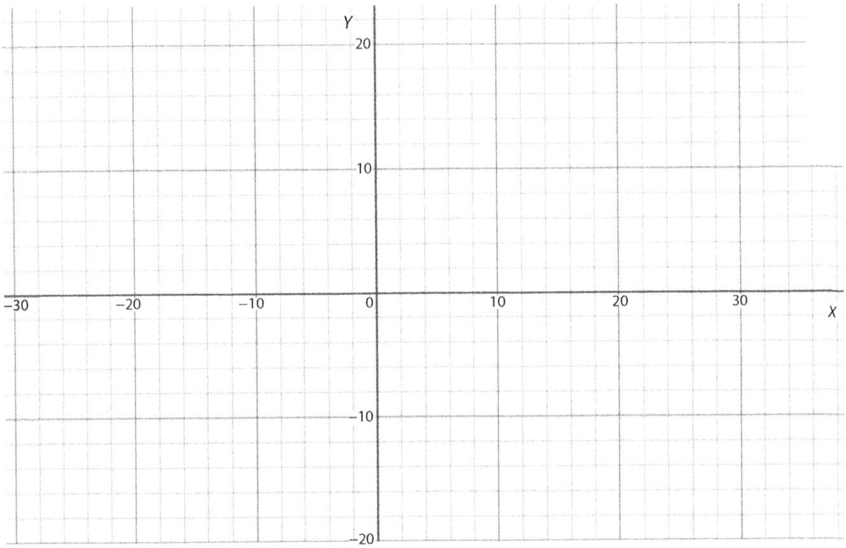

(**b**) the graph for y = 0

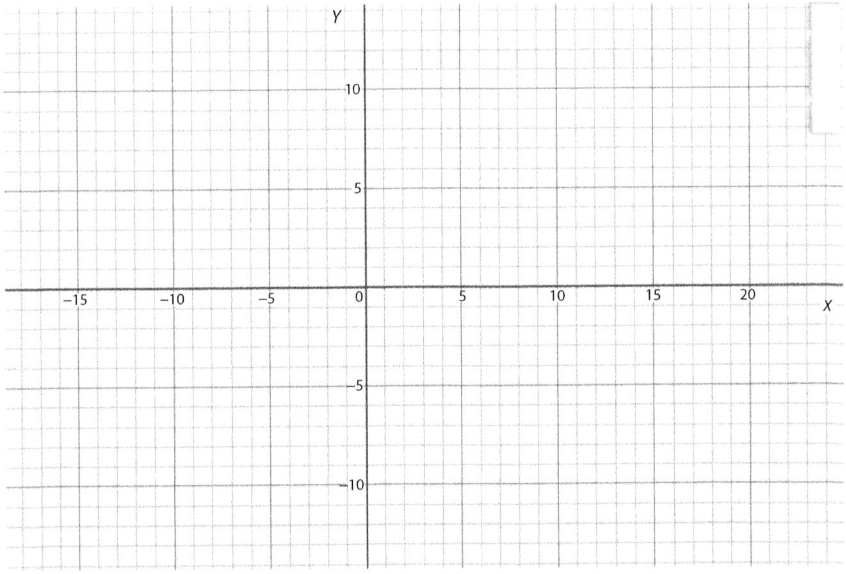

(**c**) the graph for x = 2

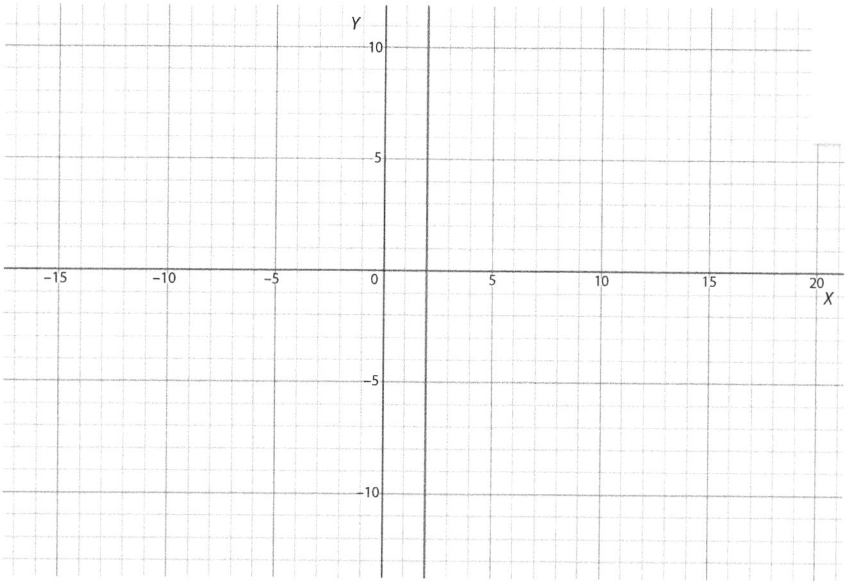

(**d**) the graph for Y = 3

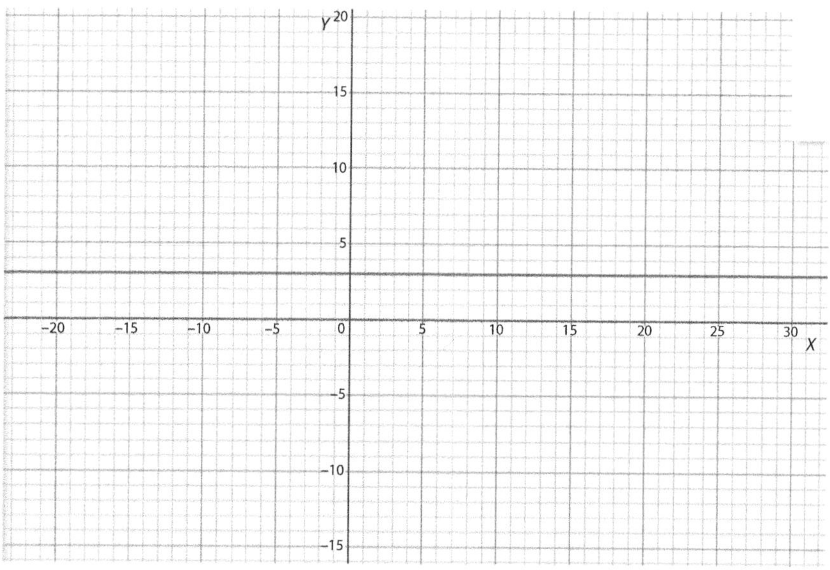

EXAMPLE 41: Draw the graph for 2x -3 =12

SOLUTION:

Solve the equation 2x -3 =12

We get x = 15/2 =7.5

We plot x=7.5 and draw the graph below.

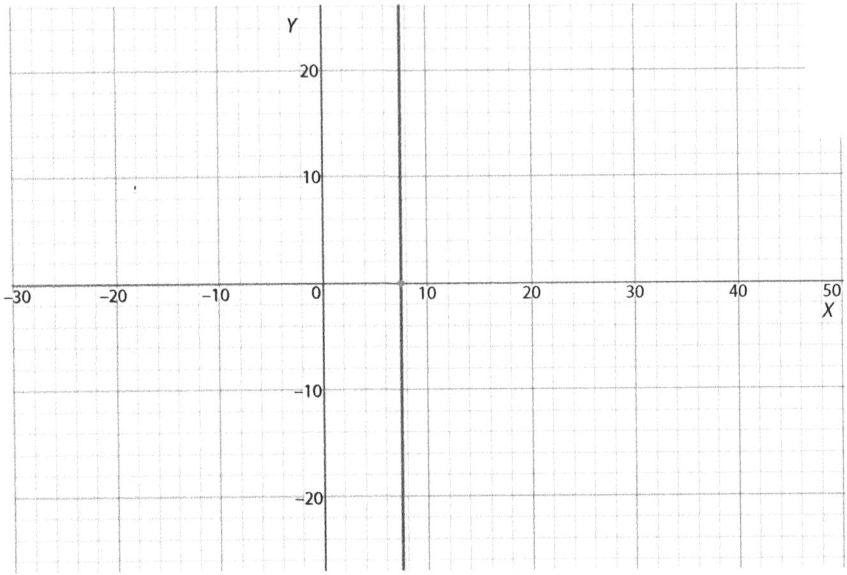

EXAMPLE 42: Draw the graph for $\frac{1}{2}x - 3 = 3$

SOLUTION:

Solve the equation $\frac{1}{2}x - 3 = 3$

We get x = 12

We plot x=12 and draw the graph below.

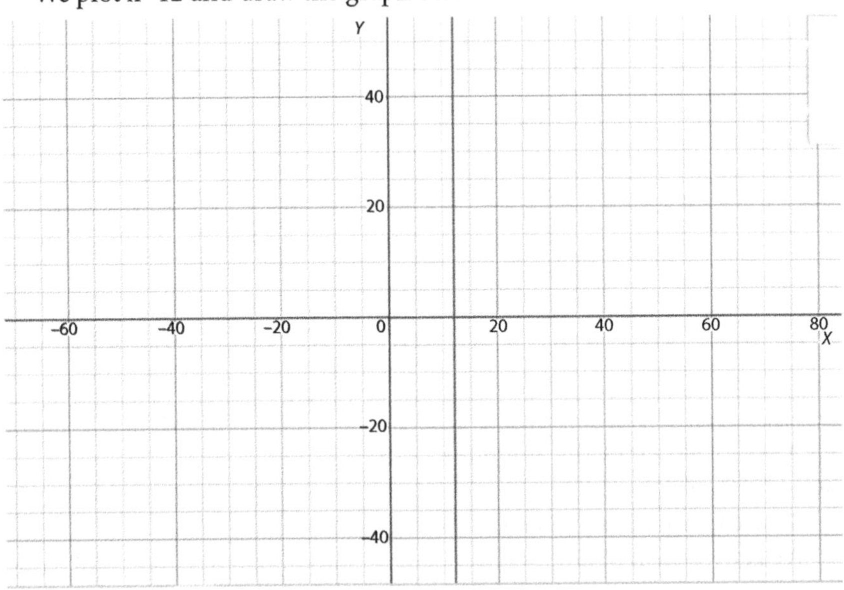

EXAMPLE 43: Graph the equation x- y = 6

SOLUTION:

Step 1: Put x =0, we get y =-6

Step 1: Put x =1, we get y =-5

Step 3: : Put x = -1, we get y = - 7

We plot the graph using the above values of y for each value of x we have considered.

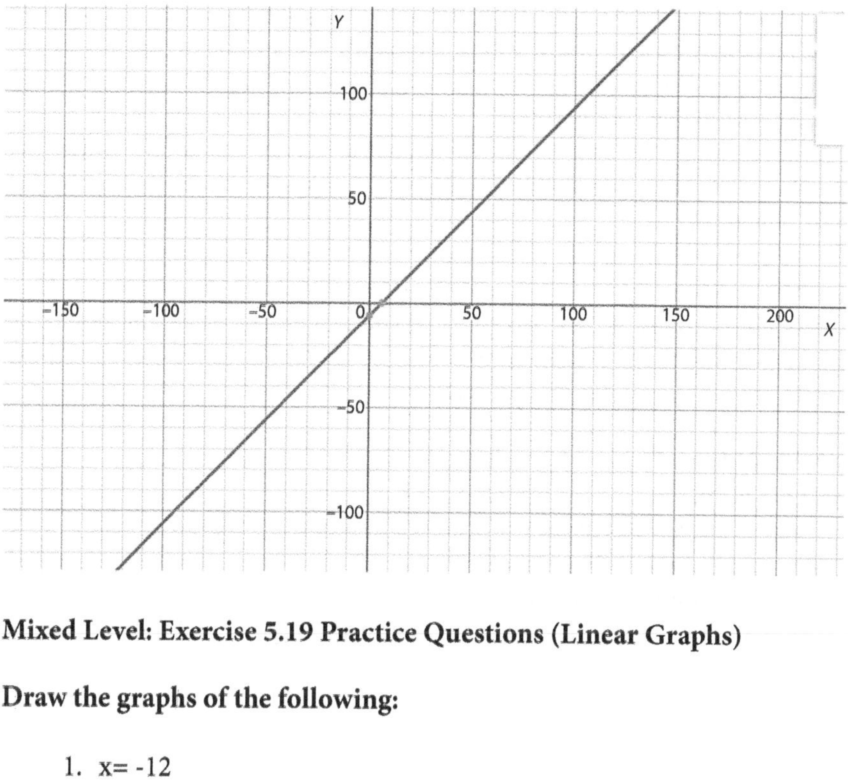

Mixed Level: Exercise 5.19 Practice Questions (Linear Graphs)

Draw the graphs of the following:

1. x= -12
2. y = -8
3. x + 5 = 20
4. y – 40 =3
5. 4x -3 =17
6. ½ x- 3 = 4
7. x + y = 10
8. 2x – y = 5
9. ½ x – 1/3y = 6
10. ¾ x – 5/6 y = 12

5.8.2 Quadratic Equations Graphs

A polynomial function of degree two that has the general formula f(x)= $f(x)=ax^2+bx+c$ is known as a quadratic function. In this case, a, b, and c are real values where $a \neq 0$. $a \neq 0$.

The graphs of all quadratic functions share a common curved shape called a parabola. We wish to add specific unique spots in the graph while graphing parabolas.

The graph's intersection with the y-axis is known as the y-intercept. The locations on the graph where it crosses the x-axis are known as the x-intercepts. The point on the graph that designates the minimum or maximum is known as the vertex.

The vertical line that passes through the vertex and along which the parabola is symmetric is the line of symmetry, sometimes known as the axis of symmetry.

We can determine the vertex and y-intercept of any parabola. We also wish to find those if the x-intercepts are present. As it is impractical to guess the x-values of these exceptional spots, we shall create methods that make locating them easier. As our study of algebra advances, a lot of these strategies will be applied frequently.

EXAMPLE 44: Draw the graph of $y = x^2$
SOLUTION:

Step 1: put x = 0, We get y = 0, so the first point is (0,0), the origin.
Step 2 : Put x = 1, Then y = 1
So the second point is (1,1)
Step 3 : put x = -1, Then y = 1, so the third point is (-1,1)
We plot the graph below using the above points.

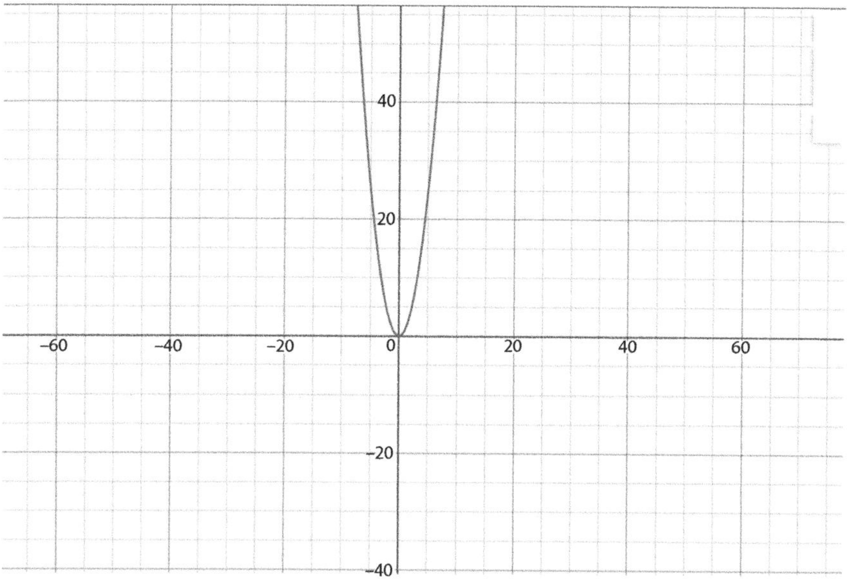

EXAMPLE 45: Draw the graph of the quadratic equation $y = x^2 + 5x + 6$
SOLUTION:

Step 1: We get the y-intercept by substituting $x = 0$

We get The point: (0, 6)

Step 2: We solve the quadratic equation: of $y = x^2 + 5x + 6$

We get the roots as -2 and -3. So -2 and -3 are the x-intercept values. (-2,-3)

Step 3 : we put $x = -5$ to get y-intercept value

So if $x = -5$ then $y = 6$

We plot the graph using the above coordinates.

(0,6), (-5,6) -2 and -3 as x-intercept values.

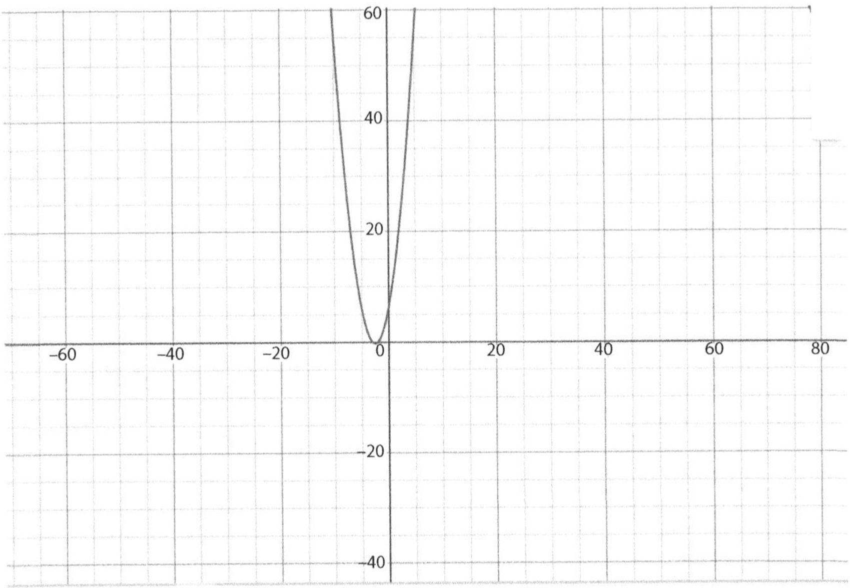

EXAMPLE 46: Draw the graph of $y = x^2 + 7x + 10$

SOLUTION:

Step 1: We get the y-intercept by substituting $x = 0$

We get the point: (0, 10)

Step 2: We solve the quadratic equation: of $y = x^2 + 7x + 10$

We get the roots as -2 and -5. So -2 and -5 are the x-intercept values.

Step 3 : we put $x = -7$ to get y-intercept value

So if $x = -7$ then $y = 10$

We plot the graph using the above coordinates.

(0, 10), (-7,10) -2 and - 5 as x-intercept values.

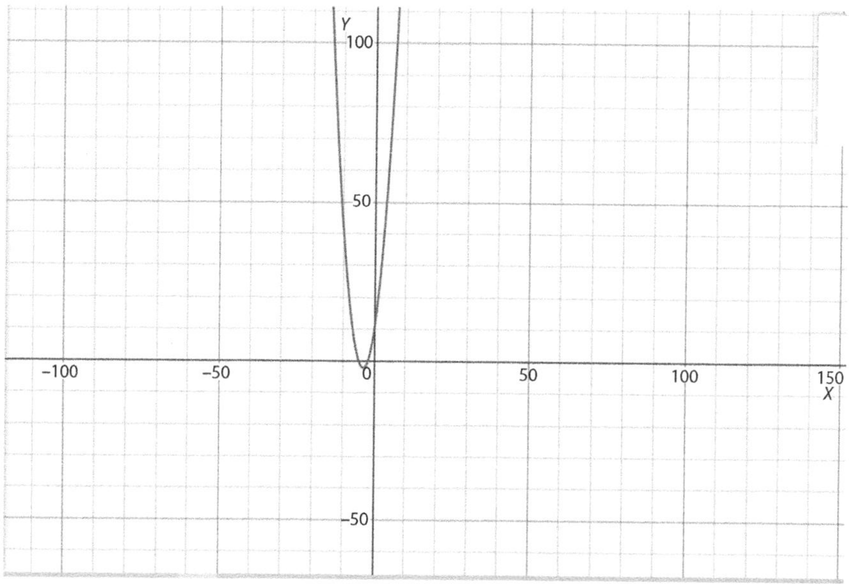

EXAMPLE 47: Draw the graph of $y = x^2 - 7x + 12$

SOLUTION:

Step 1: We get the y-intercept by substituting $x = 0$

We get the point: (0, 12)

Step 2: We solve the quadratic equation: of $y = x^2 - 7x + 12$

We get the roots as 3 and 4. So 3 and 4 are the x-intercept values.

Step 3 : we put $x = -7$ to get y-intercept value

So if $x = -7$ then $y = 12$

We plot the graph using the above coordinates.

(0, 12), (-7,12) 3 and 4 as x-intercept values.

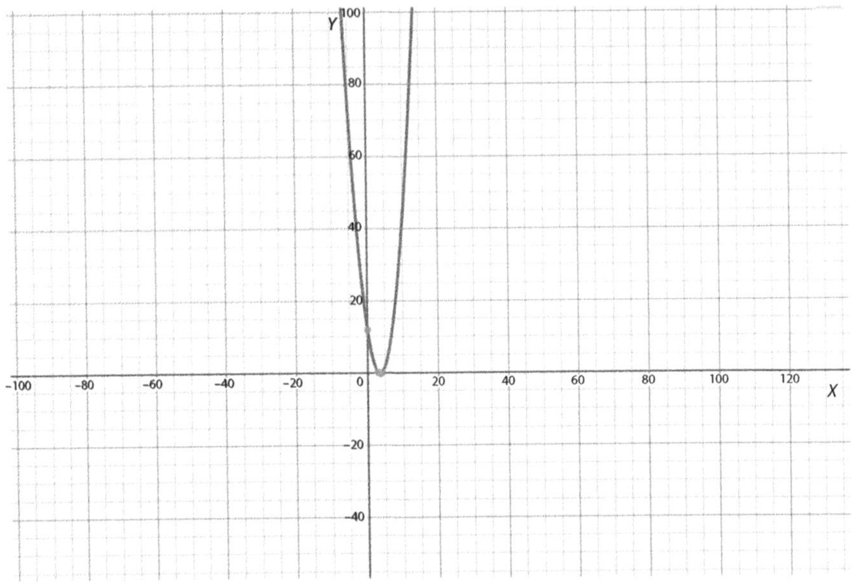

EXAMPLE 48: Draw the graph of $y = x^2 - 4x + 4$

SOLUTION:

Step 1: We get the y-intercept by substituting $x = 0$

We get the point: (0, 4)

Step 2: We solve the quadratic equation: of $y = x^2 - 4x + 4$

We get the roots as 2, 2. So 2 and 4 are the x-intercept values.

Step 3: we put $x = 2$ to get y-intercept value

So if $x = 2$ then $y = 0$

We plot the graph using the above coordinates.

(0, 4), (2,0) 2 and 2 as x-intercept values.

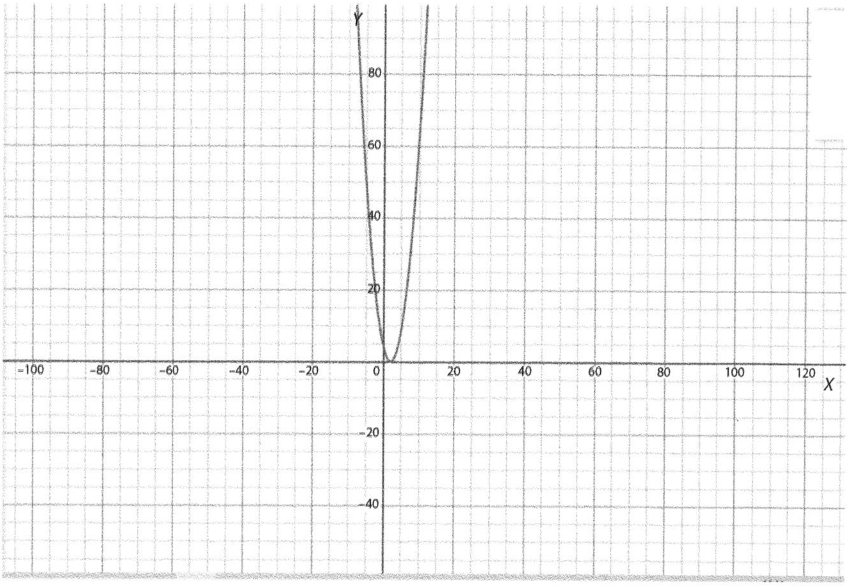

Mixed Level: Exercise 5.20 Practice Questions (Quadratic Graphs)

Draw the graph for the following quadratic equations:

1. $y = x^2 - 6x + 9$
2. $y = x^2 - 2x + 1$
3. $y = 4x^2$
4. $y = x^2 - 11x + 22$
5. $y = x^2 - 11x + 30$

5.8.3 Simultaneous Equations

Both algebraic and graphical methods can be used to solve simultaneous equations.

We must search for the intersection of two graphs to deduce solutions from them. This is known as the solution point or the singular solution for equations.

EXAMPLE 49: Draw the graph for the equations x+y =3 and x- y=1
SOLUTION: Consider the first equation x + y = 3

step 1: For x = -1, y = 4 => $P(x_1,y_1)$= (-1,4)
step 2: For x =0, y = 3 => P (x_2, y_2) = (0,3)
step 3: For x =1, y = 2 => $P(x_3,y_3)$= (1,2)

X	-1	0	1
Y	4	3	2
(x,y)	(-1,4)	(0,3)	(1,2)

Similarly, follow the same process for the equation x -y = 1

Step 1: For x = -1, y= -2 => $Q(x_1,y_1)$ = (-1,-2)
Step 2: For x = 0, y= -1 => $Q(x_2,y_2)$ = (0,-1)
Step 3: For x =1, y=0 => $Q(x_3,y_3)$ = (1,0)

X	**-1**	**0**	**1**
Y	-2	-1	0
(x,y)	(-1,-2)	(0,-1)	(1,0)

Draw the equation using the above information:

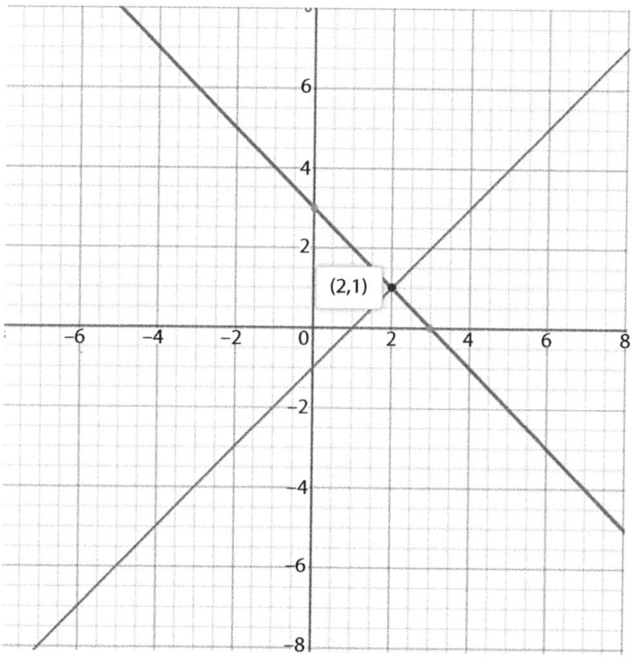

We can observe that the solution point is the coordinate point (2,1) for the equations x+y =3 and x- y=1

EXAMPLE 50: Draw the graph for the equations x+2y =3 and 3x + y =1
SOLUTION: Draw the graph using the steps shown in the previous example:

We observe that the solution point is (-1,2)

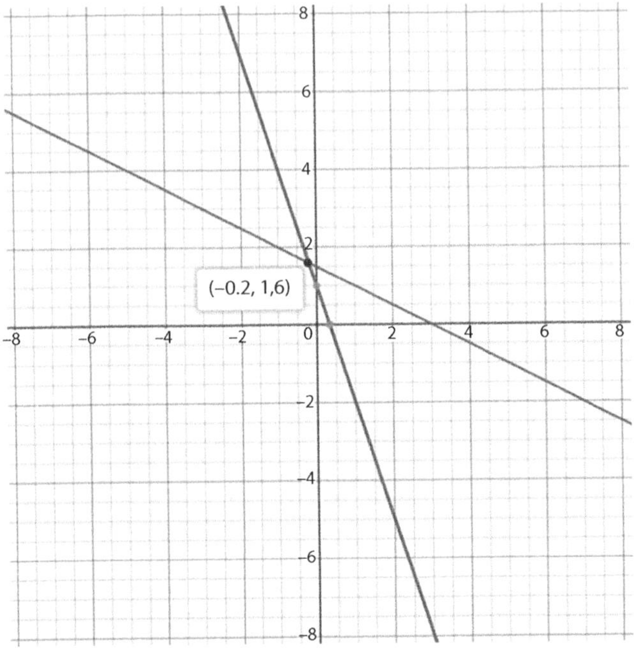

Mixed Level: Exercise 5.21 Practice Questions (Simultaneous Equations Graphs)

1. Draw the graph for the equations 2x-y =1 and 3x - 2y =1
2. Draw the graph for the equations x-y =1 and x + y = 3
3. Draw the graph for the equations x-y = 6 and x + y = 4
4. Draw the graph for the equations (x/2) -y = 4 and x + (y/4) = 4
5. Draw the graph for the equations (3x/2) – (y/6) = 1 and 3x + (7y/4) = -6

5.9 Exponential Functions

There are numerous real-world applications for the mathematical function known as the exponential function. Situations. It is mostly used to calculate investment returns, exponential decay, or exponential growth, as well as in forensics, computer science, and most life sciences. Predictions can be made using an equation that represents a real-life scenario.

How does an exponential function work?

An exponential function is an equation of form f (x) = bx, where x is a variable and b is a constant that serves as the function's base and must be greater than 0.

The exponential function, which has the formula f(x) = bx, is a significant mathematical function.

Where b is not equal to 1 and b > 0.

x can be any genuine number.

If the variable is negative, the function for -1 x 1 is undefined.

Here, "x" is a variable, and "b," the function's base, is a constant.

Exponential Function Graph

For example:

Below is the exponential function, y=2x

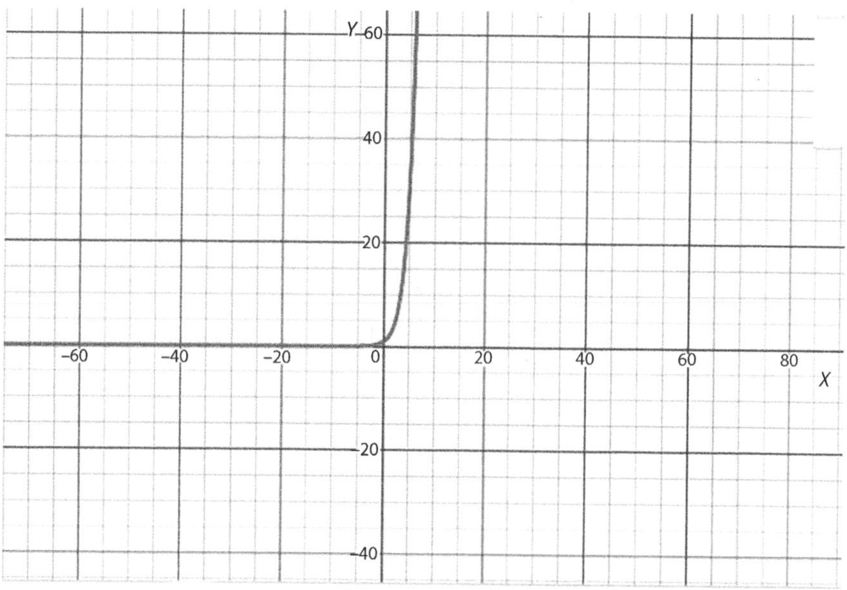

The graph passes through points (0,1).

- All real numbers make up the domain.
- y > 0 is the range.
- The graph shows an increase.

- When x gets closer to negative infinity, the graph becomes asymptotic to the x-axis.

As x gets closer to positive infinity, the graph rises unboundedly.
It is a continuous graph.
For Example: Let us consider the exponential function, y=2-x
The graph of the function y=2-x is shown below.

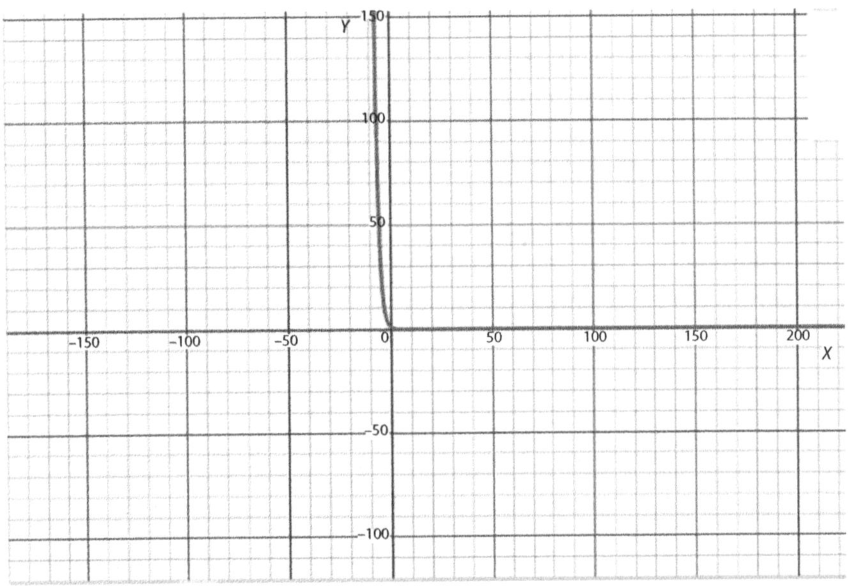

- The line intersects the location (0,1)
- The set of all real numbers is included in the domain.
- There is a range of y>0.
- When x approaches positive infinity, the line in the graph above becomes asymptotic to the x-axis, forming a declining graph.
- When x gets closer to negative infinity, the line rises without bound.
- The graph is continuous.

Mixed Level: Exercise 5.22 Practice Questions (Exponential Functions Graphs)

1. Draw the graph for the exponential function: y=3x
2. Draw the graph for the exponential function: y= (1/2) x
3. Draw the graph for the exponential function: y= (1/2)-x
4. Draw the graph for the exponential function: y= (3)-x/2
5. Draw the graph for the exponential function: y= (5/2) x

5.10 Check Your Knowledge (MCQ)

1. The playground has a 336 yd^2 space. The playground's width is five yards greater than its length. Determine the playground's length and width.
   a. length = 26 yd, width = 21 yd
   b. length = 21 yd, width = 26 yd
   c. length = 21 yd, width = 16 yd
   d. length = 16 yd, width = 21 yd

2. Solve Using the Quadratic Formula $x^2 + 4x - 40 = -8$
   a. -10 & -4
   b. -4 & 10
   c. -8 & 4
   d. 8 & -4

3. Find the zeros of this quadratic. (5b-4)(b+3)=0
   a. b=4/5, b=3
   b. b=4, b=-3
   c. b=1/5, b=3
   d. =4/5, b=-3

4. What is the slope in the equation y=2x + 3?
   a. m=2
   b. m=3
   c. m=5
   d. It cannot be Determined

5.

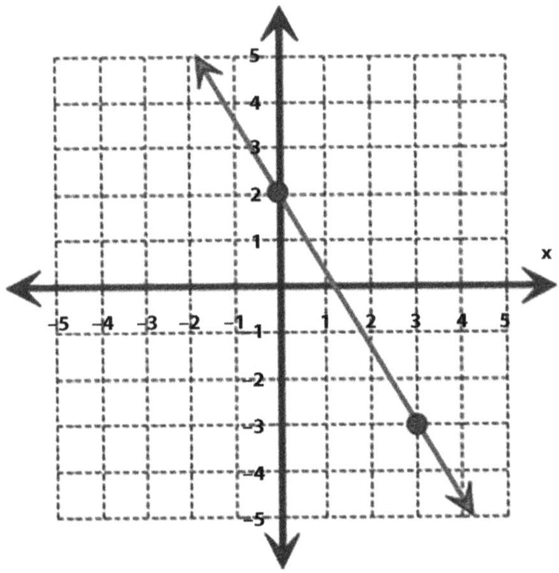

   What is the slope?
   a. m= -5/3
   b. m= -3/5
   c. m= 5/3
   d. m= 3/5

6. Which of the following are ways to solve systems of equations?
   a. Graphing
   b. Substitution
   c. Elimination
   d. All of the above

7. Solve the system of the following linear equations.
   $-4x - 4y = 0$ …..(1)
   $4x + 4y = 0$ …..(2)

a. (−6, −4)
b. An infinite number of solutions
c. (−6, 10)
d. (6, 4)

8. How many packets of chocolates can you buy for $40 if each package costs $8? (identify the correct expression)
   a. -8n = 40
   b. -8n = -40
   c. 8 - n = 40
   d. 8/n = 40

9. A rectangle's length is 6 meters longer than its width. 72 square meters make up the space. Determine the width and length.
   a. Length = 6, Width = 12
   b. Length = 18, Width = 4
   c. Length = 12, Width = 6
   d. Length = 4, Width = 6

10. Determine if (4, 1) is a solution for the system of equations.
    $y = -x + 5$....(1)
    $y = 2x - 7$....(2)
    a. yes
    b. no

5.11 Activity/Experiment

CASE 1 (Quadratic Equations)
Ahmed and Ali are extremely close friends. Each family decides to travel by automobile to a neighboring location. Ahmed's vehicle travels at x km/h, while Ali's travels at 5 km/h, faster than Amend's. Ahmed required 4 hours more than Ali did to travel the 400 kilometers.

1. How much ground will Ali's automobile travel in 2 hours?
   a) $2(x +5)$ km
   b) $(x - 5)$ km
   c) $2 (x + 10)$ km
   d) $(2x + 5)$ km

2. Which of the quadratic equations below best fits Ahmed's car's speed?
   a) $x^2 - 5x - 500 = 0$
   b) $x^2 + 4x - 400 = 0$
   c) $x^2 + 5x - 500 = 0$
   d) $x^2 - 4x + 400 = 0$

3. What is the speed of Ahmed's car?
   a) 20 km/hour
   b) 15 km/hour
   c) 25 km/hour
   d) 10 km/hour

4. How much time did it take Ali to travel 400 km?
   a) 20 hour
   b) 40 hour
   c) 25 hour
   d) 16 hour

CASE 2 (Simultaneous Equation)

The questions in an exam are either "True" or "False." For each accurate response, one mark is given; for each incorrect response, one-fourth of a mark is taken away. Some of the questions had answers that a student already knew. He guesses at the remaining questions. He received 90 points for answering 120 questions.

Question	correct answer marks	wrong answer marks
True/False	1	-0.25

1. How many questions did he properly answer if all of his guesses for answers were incorrect?
2. How many questions did he guess the answers to?
3. How many marks did he receive if he answered 80 of the questions correctly, but all of the questions he attempted by guessing were incorrect?
4. How many questions would he need to answer properly to receive 95 points if all of his guesses on the questions were incorrect?

6
Exponent and Logarithm

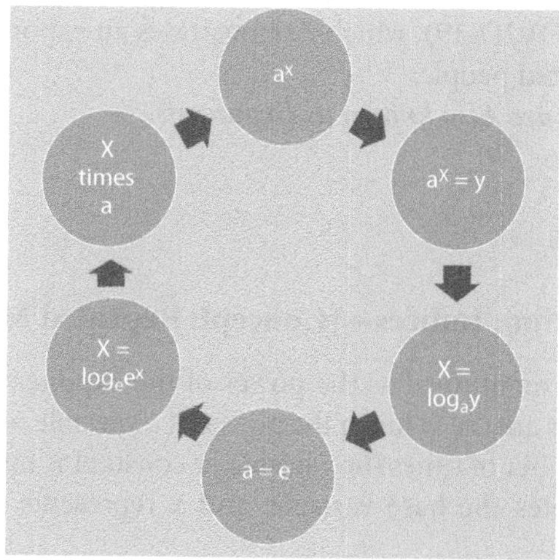

Objectives: This chapter

- *Explain concepts of Indices and exponents.*
- *Insight on Indices & Exponents Through Solved Examples.*
- *Understand Concepts of Logarithms.*
- *Insight on Indices & Exponents Through Solved Examples.*
- *Group Project work on Applications of Indices, Exponents, and logarithms*

6.1 Introduction

HISTORY-The basic idea of indices originated in the 9th century when Persian Mathematician Musa explored the square of a number and later the cube in the 15th century with different notifications. The words "exponent" and "indices" were first used in literature in 1544 and 1696, respectively.

Ritu Shrivastava, Ramakant Bhardwaj and Satyendra Narayan. Mathematics for Engineers, (249–266)

Exponential notation reached maturity in the 17th century and began to be used in problems by mathematicians worldwide.

Exponents or Indices have many applications. The most popular applications are denoting population growth, the decay of chemicals in subsequent years, and many other fields. One of the most recent examples of exponents is predicting a pattern for the spread of the pandemic New Coronavirus (COVID-19), which demonstrates an exponential rise in the number of infected people.

Note: Indices are Also known as Exponents.

6.2 Indices

6.2.1 Definition: Indices—[Concept: Repeated Multiplication]

Indices are also mentioned as the power of the exponent of a variable or constant. Indices are the value in the power of a variable or constant, which denotes the number of times the variable or constant is multiplied by itself.

In a^x, a denotes the base variable, and x represents the index of the variable.

The generalized form of indices is $a^n = a \times a \times \ldots a$ (n times), 'a' read as 'a' to the power 'n.'

note: 'INDEX,' also known as 'POWER.'

for example: $3^4 = 3 \times 3 \times 3 \times 3$,

Here 3 is called the **base,** and 4 is the **index** of the Base.

Important Note: The Index value may be zero, a positive integer, a negative integer, or a fraction. Depending upon the value and for calculating two terms containing indices, some rules are framed for easier algebraic calculation.

Zero index: An example of 6^0, here 6 is the base & index is ZERO.

Positive index: An example of a is 3^5; here, 3 is the base & index is 5. Since 5 is a positive number, the index takes a positive value.

Negative index: An example of 5^{-6} here 5 is the base & index is -6. Since -6 is a Negative number, the index takes a Negative value.

6.2.2 Rules of Indices

RULE 1: Product of Powers Rule (Concept – Addition)
The multiplication of two terms with the same base and different indices is equal to the same base with the sum of the different indexes or power.

$\mathbf{a^x \times a^y = a^{x+y}}$ [Concept – x times a multiplied by y times a = $(x + y)$ times a]

Example: $3^5 \times 3^3 = 3^{(5+3)} = 3^8$
Example: $a^8 \times a^7 = a^{(8+7)} = a^{15}$
Example: $5^{-6} \times 5^8 = 5^{(-6+8)} = 5^2$

RULE 2: Quotient of Powers Rule (Concept – Subtraction)
The division of two terms with the same base and different indices is equal to the same base with a difference of powers.

$\mathbf{a^x \div a^y = a^{x-y}}$ [Concept: x times a divided by y times a will lead to cancellation of a, y times from x times]

Example: $6^5 \div 6^3 = 6 \times 6 \times \cancel{6} \times \cancel{6} \times \cancel{6} \div \cancel{6} \times \cancel{6} \times \cancel{6} = 6^{(5-3)} = 6^2$
Example: $b^6 \div b^3 = b \times b \times b \times \cancel{b} \times \cancel{b} \times \cancel{b} \div \cancel{b} \times \cancel{b} \times \cancel{b} = b^{(6-3)} = b^3$
Example: $5^{-5} \div 5^3 = 5^{(-5-3)} = 5^{-8}$

RULE 3: Power of Power Rule (Concept – Multiplication)
The term, a variable or constant with a power (indices) increased to another power, is equivalent to the variable with the multiplication of the powers.

$\mathbf{(a^x)^y = a^{xy}}$ [Concept: y times a^x is xy times a]
Example: $(5^5)^3 = 5^{(5\times3)} = 5^{15}$

Example: $(c^6)^3 = c^{(6\times3)} = c^{18}$

RULE 4: Zero Index (Concept –Division of Powers Rule)
The resultant value equals one if the term has zero power or index.

$\mathbf{a^0 = 1}$ **(Note base a cannot be zero)** [Concept: Anything divided by itself is 1, so $1 = x^3 \div x^3 = x^{(3-3)} = x^0$]

Example: $5^0 = 1$
Example: $c^0 = 1$

RULE 5: Negative Index (Concept – Division of Powers Rule)
If the term has negative power or index, then the resultant value equals the inverse of the positive power.

$\mathbf{a^{-x} = 1/a^x}$ [Concept: $1/a^x = a^0/a^x = a^{(0-x)} = a^{-x}$]

Example: $5^{-1} = 1/5$
Example: $x^{-6} = 1/x^6$
Example: $(2/3)^{-8} = (3/2)^8$
Example: $2^{-5}/2^3 = 1/2^5 \times 2^3 = 1/2^8$
Example: $6^{-3} \times 6^{-6} = 1/6^3 \times 1/6^6 = 1/6^{(3+6)} = 1/6^9$
Example: $a^5 \times a^{-3} = a^5 \times 1/a^3 = a^2$

RULE 6: Fractional Index
Suppose the term has fractional power or index. In that case, the numerator of the fraction denotes the usual power. The denominator indicates the type of the root and is equal to the denominator index root of the numerator power of the variable or constant.

$$\sqrt{\mathbf{a}} = \mathbf{a}^{1/2} \textbf{ and } \mathbf{a}^{x/y} = \sqrt[y]{\mathbf{a}^x} \text{ [Concept: } a^{x/y} = (a^x)^{1/y} = \sqrt[y]{a^x}$$

Example: $5^{5/2} = (5^2)^{1/2} = \sqrt[2]{5^5}$
Example: $a^{1/3} = \sqrt[3]{a}$
Example: $5^{1/2} \times 5^{2/5} = 5^{(1/2+2/5)} = 5^{(5+4)/10} = 5^{9/10}$

RULE 7: Power of a Product
The term which is a product of two variables or constants with a whole power or index is equal to the product of two variables or constants with individual powers.

$\mathbf{(ab)^x = a^x b^x}$
Example: $(ab)^3 = a^3b^3$
Example: $(2 \times 3)^6 = 2^6 \times 3^6$
Example: $8^6 \times 4^6 = (8 \times 4)^6$
Example: $(6 \times 3)^{-6} = 6^{-6} \times 3^{-6}$
Example: $(5 \times 4)^{2/3} = 5^{2/3} \times 4^{2/3}$

RULE 8: Power of a Quotient
The term, which is a quotient of two variables or constants with a whole power or index, is equal to the quotient of two variables or constants with individual powers.

$(a/b)^x = a^x / b^x$
Example: $(c/d)^3 = c^3/ d^3$
Example: $(2/3)^6 = 2^6 / 3^6$
Example: $(p/q)^{2/3} = p^{2/3}/q^{2/3}$
Example: $(5/6)^{-9} = 5^{-9}/6^{-9}$
Example: $(8/a)^0 = 8^0/a^0 = 1$
Example: $(8\times3/6\times5)^6 = (8^6 \times 3^6)/(6^6 \times5^6)$

SUMMARY OF LAW OF INDICES

Index rules	Definition	Notation
Index Law 1 Product of Powers	The product of terms with the same base implies adding the indices	$a^x \times a^y = a^{x+y}$
Index Law 2 Quotient of Powers	Division of terms with the same base implies subtracting the indices	$a^x \div a^y = a^{x-y}$
Index Law 3 Power of Power	Raising a term with power to power implies multiplying the indices	$(a^x)^y = a^{xy}$
Index Law 4 Power of Product	The power of the product term is the product of the power terms	$(ab)^x = a^x b^x$
Index Law 5 Power of Quotient	Power of the division term is the division of the power terms	$(a/b)^x = a^x/ b^x$
Zero Index	A term with a zero index is one	$a^0 = 1$
Negative Index	A term with a negative index is the inverse term with a positive index	$a^{-x} = 1/a^x$
Fractional Index	A term with a fractional index is the denominator index root of the numerator power of the base	$a^{x/y} = \sqrt[y]{a^x}^{\,y}$

Level 1: Exercise 6.1 Practice Questions

1. What is the Exponent in 3^4?
2. Express $3 \times 3 \times 3 \times 3$ in Exponent form.
3. Find the value of 2^3.
4. The value of 5^0 is equal to?
5. The Exponent form of $6 \times 6 \times 6 \times 6 \times 6$ is?
6. The value of $2^3 + 3^2$ is?
7. $9 \times 9 \times 9 \times 8 \times 8 \times 8$ can be expressed in Exponent form as ______.
8. Find the product of $2^2 \times 2^3$.
9. The Minimized Exponent form of $2^3 \times 2^4 \times 3^5 \times 3^8$ is _____.
10. Express the value of $2^7 / 2^3$ as an Exponent of 2.
11. The value of $(2^3)^2$ is the same as the value of ______?
12. Find the value of 2^{-1}.
13. Express $(-2) \times (-2) \times (-2)$ in Exponent form.
14. The product value of $(-1) \times (-1) \times (-1) \times (-1)$ is equal to _____?
15. Find the value of $(32)^2$.
16. Convert $(2^3 \times 2^5) / 2^2$ into an exponent of 2.
17. Convert the exponent $(-2/3)^{-1}$ into positive power.
18. Find the decimal or numeric value of $(2^3)^2$.
19. What is the index of 2 when 2^{10} is divided by 2^5?

Level 2: Exercise 6.2 Practice Questions

1. The Exponent $(2/3)^2 \times (2/3)^5 \times (2/3)^6$ can be reduced as _______.
2. Find the numeral value of $2^0 + 3^0 + 5^0$.
3. Let n be any number; the value of n^0 always equals ____.
4. Find the reduced positive exponent of $2^{-1} + 3^{-1}$.
5. Express $3^5 \times 3^8 \times 3^{-2}$ in the simplified form.
6. If $2^n = 16$, then find the value of n.
7. The equivalent expression of $2^2 \times 3^2$ is equal to _____.
8. The simplified form of $3^{-1} \times 3^{-2} \times 3^{-3}$ equals _____.
9. Find the numeral value of $(3^2)^2 + (2^3)^2$.
10. Simplify $\{2^{-1} \times 5^{-1}\}^2$.
11. Simplify: $\{2^0 \times 5^0 \times 2^{-3}\}$.
12. Express each of the following as a simplified positive Exponent.
    a. 2^{-1}
    b. $(2/3)^{-1}$
    c. $(2/7)^{-2} \times (7/2)^{-3}$

13. Evaluate $\frac{8\times3}{2^3}$

14. Simplify $\frac{25\times m^{-3}}{5^{-2}\times10\times m^{-2}}$

15. Simplify $\frac{25\times3^{-3}\times5^{-3}}{27\times5^4}$

16. Find the value of n if $(3/7)^5\times(3/7)^{-10}=(3/7)^{5n}$
17. If $x=(5/2)^2\times(2/5)^{-5}$, find the value of x.
18. Find the value of $2^{17}\div2^{15}$.
19. Find the value of $3^{18}\times3^{19}\div3^3$.
20. What should be multiplied to $(5/3)^{-2}$ so the product will be $(13/3)^{-1}$?

Level 3: Exercise 6.3 Practice Questions

1. Find x, so that $(7/3)^{-5}\times(7/3)^{-12}\times(3/7)^{-12}=(7/3)^{3x}$.
2. Which integer should be used to divide $(-15)^{-3}$ so that the result is $(3/2)^{-1}$?
3. Simplify $\frac{625\times3^{-9}\times10^{-6}}{5^{-9}\times6^{-3}}$
4. The Sun and Earth have dimensions of 1.4 × 109 meters and 1.295 × 107 meters. Calculate how many times larger the Sun's diameter is than the Earth's.
5. A plant cell is 0.00001275 times smaller than a red blood cell, which is 0.000007 millimeters. The two are comparable.
6. Simplify $\frac{64\times512^{\frac{1}{3}}\times8^{12}\times2^{18}}{256}$
7. Find the value of $16\times(256)^{-3/4}\div(216)^{1/3}$?
8. If $\left(\sqrt{3}\right)^{m+n}=9$ and $\left(\sqrt{27}\right)^{m-n}=81$. Find the value of 2mn.
9. If $3^{(x+y)}=27$ and $4^{(x-y)}=16$, find the value of x/y.
10. If $(216x)^{12}(6x)^{32}=60^{20x}$, then find the value of x?
11. Simplify $64^1\times64^2\times64^3\times\ldots\times64^{50}$
12. If $x=((64)^{3/2}(128)^{2/7})$, then find the value of x^{-3}.
13. Simplify $\frac{x^2\times x^3\times x^4\times x^{-8}}{x^{12}}$

14. Simplify $\dfrac{x^{m} \times y^{m} \times x^{-3m}}{(x^{2})^{5m} \times (y^{3})^{2m}}$
15. Find the value of $(1^0 + 2^0 + 3^0 + 4^0 + \ldots. + 100^0)^{1/2}$?
16. Simplify $\dfrac{(2^{2})^{3m} \times (4^{3})^{2n}}{(16^{2})^{2m}}$
17. Find the value of $(0.000216)^{5/3} + (0.0256)^{3/4}$
18. Given that If $\left(\sqrt{3}\right)^{n} = \dfrac{3^{x}}{(27)^{y}}$ Express n in terms of x and y.
19. Find the value of $\dfrac{1}{1+a^{n-m}} + \dfrac{1}{1+a^{m-n}}$.
20. If $x = 3 + 2\sqrt{23} + 2\sqrt{2}$, then fins the value of $\sqrt{x} - \dfrac{1}{\sqrt{x}}$

6.3 Logarithms

The Concept of Logarithm initially originated in the ancient Babylonian period when records in baked clay tablets contained successive powers of whole numbers. This leads to the question, "To what power must a certain number be raised to yield a given number," which is equivalent to the modern logarithm's definition.

Logarithms were invented in two approaches independently by John Napier and Scotsman in 1620 with an algebraic approach and by Joost Burgi, a Swiss, in 1620, with the geometric approach. Both men had not used the logarithmic base concept but simplified the tedious mathematical calculation of very large and very small numbers from multiplication and division to addition and subtraction.

The possibility of defining logarithms occurred as **John Wallis recognized exponents** in 1685 and Johann Bernoulli in 1694.

Logarithms are helpful in many fields, from finance to astronomy.

6.3.1 Definition

A logarithm is an exponent or power 'x' to which a number a, called the base, must be raised to yield a specific number n. Logarithms also known as inverse function of exponentiation

Generally, $\mathbf{a^x = n}$ **is same as** $\mathbf{x = \log_a n}$**,** here, x is the logarithm of n to the Base a.

Note: Base a & n is always positive, i.e., a > 0, n > 0, and Base b ≠ 1.

Indices and logarithms are inversely related to each other. Let us see by some examples how they are related and how to differentiate the usage of indices and logarithms by notations.

To convert from exponential form to logarithmic form, we use the relationship.

$$(a^x = n) = (x = \log_a n)$$

For example:

Exponential form	Logarithm form
$3^3 = 27$	$3 = \log_3 27$
$2^5 = 32$	$5 = \log_5 32$
$6^2 = 36$	$2 = \log_6 36$
$7^3 = 343$	$3 = \log_7 343$
$10^3 = 1000$	$3 = \log_{10} 1000$
Try These	
$2^6 = 64$	
	$\log_6 216 = 3$
$8^2 = 64$	
	$\log_9 729 = 3$
$5^3 = 125$	

6.3.2 Types of Logarithms [Depending on Base]

Depending on the base, the most commonly used logarithms are classified into two types:

1. The **Common Logarithm** is a logarithm with **base 10**, denoted by **$\log_{10}$ or log.**
2. The **Natural Logarithm** is a logarithm with the **base 'e'** denoted by **$\log_e$ or ln.**

[*Note: Here, e is the exponential constant = 2.718281*]

As we discussed above, indices and logarithms are related inversely. We can correlate the rules of indices and the rules of logarithms.

6.3.3 Rules of Logarithm

RULE 1 (Product Rule): The First Law of Logarithms

The sum of the logarithms of the two numbers is equal to the logarithm of the product of two numbers (n, m) with the same Base a.

$$\boxed{\log_a(n \times m) = \log_a n + \log_a m}$$

Note: For logarithmic calculation, the base must be the same for all terms.

Example: $\log_{10}16 + \log_{10}4 = \log_{10}(16\times4) = \log_{10}64$

Example: $\log_3(6\times3) = \log_3 6 + \log_3 3$

RULE 2 (Power Rule): The Second Law of Logarithms

A number's logarithm is defined as the product of its power and logarithm.

$$\boxed{\log_a(n^m) = m \log_a n}$$

The corollary of Rule 2:

1. If $m = \frac{1}{x}$, $\log_a n^{1/x} = \frac{1}{x} \times \log_a n = \frac{\log_a n}{x}$

Example: $\log_3 5^6 = 6 \times \log_3 5$

Example: $\log_6 8^3 = 3 \times \log_6 8$

Example: $\log_2 4^{1/2} = \frac{\log_2 4}{2}$

Example: $\frac{\log_3 27}{3} = \log_3 27^{1/3} = \log_3 3 = 1$

RULE 3 (Quotient Rule): The Third Law of Logarithms

The logarithm of the quotient(division) of two numbers (n, m) with the same base a is equal to the **difference** between the two numbers' logarithm.

$$\boxed{\log_a \frac{n}{m} = \log_a n - \log_a m}$$

Example: $\log_{10}\frac{12}{13} = \log_{10}12 - \log_{10}13$

Example: $\log_{10}12 - \log_{10}2 = \log_{10}\frac{12}{2} = 6$

Example: $\log_{6}1 = 0$

Example: $\log_{10}10 = 1$

Important Results:

1. $\mathbf{\log_a 1 = 0}$
   From the indices rule, we know that $a^0 = 1$
   By logarithm definition, $0 = \log_a 1$
2. $\mathbf{\log_a a = 1}$
   We know that $a^1 = a$
   By logarithm definition, $1 = \log_a a$
3. **Change of Base Rule:** $\boldsymbol{\log_a b = \frac{\log_c b}{\log_c a}}$
4. **The inverse of logarithm:** $\mathbf{\log_a b \times \log_b a = 1}$

6.3.4 Characteristics and Mantissa of Logarithms

The logarithm value for all numbers for all bases can be extracted from logarithmic tables

Every logarithm value has two parts Characteristic and Mantissa.

Characteristic: The integer part of the common logarithm is called the characteristic.

Mantissa: The non-negative decimal part of the common logarithm is called Mantissa

Note: For negative logarithm value, Add & Subtract 1, (1–negative decimal part) gives non-negative decimal part Mantissa, (–1– negative integer part) gives characteristic.

Example: Find the characteristic and Mantissa for log 39.2.

log 39.2 = 1.5933, Here 1 is the characteristic and .5933 is the Mantissa

Example: Find the characteristics and Mantissa for – 15.78.

Suppose log x = - 15.78, Here log x = – 15.78 + 1 – 1 = – 15 – 0.78 + 1 – 1
= (– 15 – 1) + (1 – 0.78) = – 16 + 0.22

Hence -16 is the characteristic and 0.22 is the Mantissa.

INTERESTING APPLICATION OF LOGARITHMS

Can we find the number of digits of 2^{2020}????

Yes, it can be possible effortlessly by logarithms

The amount of digits in a given number is one more than the number's characteristic logarithm value.

Example: Find the number of digits of log 2^{10}, when log 2 = 0.3010.

Solution: log $2^{10} = 10 \times \log 2 = 10 \times 0.3010 = 3.010$, here characteristic is 3, and hence the number of digits of 2^{10} is 4.

6.3.5 Properties of Logarithms

Logarithm rules	**Algebraic notations**
Product Rule	$\log_a(nm) = \log_a n + \log_a m$
Power Rule	$\log_a n^m = m \times \log_a n$
Quotient Rule	$\log_a \frac{n}{m} = \log_a n - \log_a m$
Change of Base Rule	$\log_a b = \frac{\log_c b}{\log_c a}$
Inverse Logarithm	$\log_a b = \frac{1}{\log_b a}$
Logarithm of 1	$\log_a 1 = 0$
Logarithm of Same Base & Number	$\log_a a = 1$

Level 1: Exercise 6.4 Practice Questions

1. Find the value of $\log_2 1024$.
2. Calculate $\log_2 16$.
3. Find the value of $\log_{18} 324$.
4. Find the value of $\log_2 64 - \log_2 4$.
5. The logarithmic form of $a^b = c$ is?

6. The logarithmic form of $19^2 = 361$ is?
7. Convert log (81×25) into an additional form.
8. Expand $\log \frac{125}{16}$ into subtraction form.
9. Find the value of $\log 10^3$.
10. Calculate the value for $\log_{10} 100$.
11. Find the expanded addition form of $\log_2 10$.
12. Calculate the value for $\log_7 \sqrt{49}$.
13. Calculate the value for $\log_3 \frac{1}{27}$.
14. The value of $\log_6 1$ =?
15. Express $\log_3 729$ in indices form and find the value.
16. The Exponential form of $\log_4 64 = 3$ is.
17. Which number has no multiplicative inverse?
18. Expand $\log_5 15$ as the addition of two logarithms.
19. The value of $\log_{10} 2 + \log_{10} 5$ is =?
20. The value of $\log_b a \times \log_a b$ is =?

Level 2: Exercise 6.5 Practice Questions

1. The value of $2 \log 3 + 3 \log 5 - 5 \log 2$ =?
2. If $3 \log(x+3) = \log 27$, then the value of x is=?
3. Find the expanded form of $\log \frac{x^2 y^3 z^4}{w^5}$.
4. The exponential form of $\log_{10} 0.001 = -3$ is?
5. The logarithm of 0.0625 to the base 2 is?
6. If $\log 27 = 1.431$, the value of $\log 9$ is?
7. If $\log_{10} 2 = 0.3010$, Find the value of $\log_{10} 1600$?
8. If $\log_8 (1/2)x = (10/3)$, find the value of x.
9. Find the value of $\frac{\log \sqrt{6}}{\log 6}$.
10. If $\log \frac{a}{b} + \log \frac{b}{a} = \log a + b)$, then find the value of (a+b).
11. Find the value of $\frac{1}{\log_3 60} + \frac{1}{\log_4 60} + \frac{1}{\log_5 60}$.
12. If $\log_x \frac{9}{16} = -\frac{1}{2}$, then x is equal to?
13. If $\log_x y = 100$ and $\log_2 x = 10$, then the value of y is?

14. If $\log_2 [\log_3 (\log_2 x)] = 1$, then x is equal to?
15. Find the value of the expression $\log\frac{9}{14} - \log\frac{15}{16} + \log\frac{35}{24}$.
16. If $\log_5(x^2+x) - \log_5(x+1) = 2$, then find the value of x.
17. If $\log_{10}7 = a$, then $\log_{10}\frac{1}{70}$ is equal to?
18. If $a^x = b^y$, then find the value $\frac{\log a}{\log b}$.
19. If $\log_{10}a = p$, $\log_{10}b = q$, then what is the value of $\log_{10}(a^p b^q)$?
20. If log 2 = 0.3010 and log 3 = 0.4771, then the value of $\log_5 512$=?

Level 3: Exercise 6.6 Practice Questions

1. If log 2 = 0.3010, find the number of digits in $(125)^{25}$.
2. Find the value of the logarithm expression $\log\frac{a^n}{b^n} + \frac{b^n}{c^n} + \frac{c^n}{a^n}$.
3. Find x, if log x = log 1.5 + log 12.
4. If $\log_{10}a = b$, find the value of 10^{3b} in terms of a.
5. Find x, if $\log x = \log 7.2 - \log 2.4$.
6. Solve for x, $\log (x^2-6x+6) = 0$.
7. Find the value of x, if $5^x = 10$.
8. If log 3 = 0.4771, find $\log(.81)^2 \times \log\left(\frac{27}{10}\right)^{\frac{2}{3}} / \log 9$
9. Find the biggest among the two logarithmic terms $\log_{11}1650$, $\log_{13}1950$?
10. Find the value of $\frac{\log\sqrt{6}}{\log\sqrt[3]{6}}$.
11. If $\left(\frac{21}{10}\right)^x = 2$ express the value of x in logarithms.
12. If log 2 = 0.3010 and log 3 = 0.4771, find the number of digits of $(648)^5$.
13. If $\log_4 5 = a$ and $\log_5 6 = b$, find the $\log_3 2$?
14. If $\log_{30} 3 = x$ and $\log_{30} 5 = y$, find the $\log_8 30$.
15. What is the value of $\frac{1}{\log_2 n} + \frac{1}{\log_3 n} + \frac{1}{\log_4 n} + \cdots \frac{1}{\log_{40} n}$?
16. If log 2 = 0.3010, then the number of digits of 2^{64} is?
17. Find the value of $\log_{15} 3375 \times \log_4 1024$.
18. Find the number of digits of 8^{25}.

19. The relationship between the population p(t) growth and time (t) in years is given by the equation $p(t) = 2^t$. Determine how long it would take the population to reach 100,000?
20. How long will it take to have fewer than 1,000 cells in a biology experiment where 5,000,000 cells and 45%

($e^b = 0.55$; b - decay factor) of the cells die every minute? (The relationship equation is $f(t) = f(0)\, e^{bt}$)

6.4 Check Your Knowledge (MCQ)

1. Find the index of 5^3 is ____?
   a. 5
   b. 3
   c. 125
   d. None of these

2. The value of $(-2)^2 \times (-2^3)$ is?
   a. 32
   b. 16
   c. -16
   d. None of these

3. The index value of 6 in the expression $6^3 \div 6^2 \times 6^1$ is?
   a. 1
   b. 2
   c. 0
   d. None of these

4. The exponential form of $5 \times 6 \times 5 \times 6 \times 5$ is?
   a. $5^3 \times 6^2$
   b. $3^5 \times 2^6$
   c. $5^3 \times 2^6$
   d. $3^5 \times 6^2$

5. Find the value of $5^0 + 1^9 + 0^8$ is equal to?
   a. 2
   b. 6
   c. 14
   d. None of these

6. Solve for x, if $3x - 3^2 = 0$?
   a. 1
   b. 9
   c. 3
   d. None of these

7. Find the reduced form of the expression $(-1/3) \times (3^9/2^6) \times (2^8/3^7)$.
   a. 12
   b. -12
   c. 36
   d. -36

8. If $6^x = 216$, find the value of x?
   a. 2
   b. 3
   c. 4
   d. None of these

9. Evaluate $4^4 \times 8^{-2} \times 2^3 \times 2^{-1} \times 4^{-2}$ =?
   a. 2
   b. 4
   c. 8
   d. None of these

10. The positive exponent of $8^{-9} + 7^4 - 7^2$ is?
    a. $(1 \div 8^9) + 7^2$
    b. $(1 \div 8^9) + (7^2 \times 4^8)$
    c. $8^9 + (7^2 \times 4^8)$
    d. None of these

11. The number to be multiplied with $(8/9)^7$ gives $(4/9)^2$.
    a. $9^5 \times 2^{-17}$
    b. $9^5 \times 2^{17}$
    c. $9^5 \times 2^{-16}$
    d. $9^5 \times 2^{16}$

12. Estimate the value of the expression $64^{1/3} + 2^2 + 81^{1/2}$?
   a. 27
   b. 16
   c. 17
   d. None of these

13. Find the value of $(0.000125)^{1/3} + (0.0016)^{1/2}$ is?
   a. 0.0045
   b. 0.09
   c. 0.0009
   d. None of these

14. If $x = 1^+$, then find the value of x^{-1}?
   a. 1-
   b. -1+
   c. -1-
   d. None of these

15. The logarithmic form of $8^3 = 512$ is?
   a. $\log_8 512 = 3$
   b. $\log_3 8 = 512$
   c. $\log_{512} 8 = 3$
   d. None of these

16. Estimate the value of the expression $\log_2 2 + \log_2 1 + \log_2 8$?
   a. 2
   b. 3
   c. 4
   d. None of these

17. Find the value of $3 \log 5 + 2 \log 6 - 2 \log 3$ =?
   a. log 125
   b. log 250
   c. log 500
   d. None of these

18. How many digits are in 2^{15}, if $\log 2 = 0.3010$?
   a. 2
   b. 3
   c. 4
   d. 5

19. Solve for x, if $\log_6(x^2+4x+5) = 0$?
   a. 2,2
   b. 2,3
   c. 3,3
   d. None of these

20. Find the value of $\log_{216}6 + \log_9 243$?
   a. 17/6
   b. 8/3
   c. 14/3
   d. None of these

7
Trigonometry

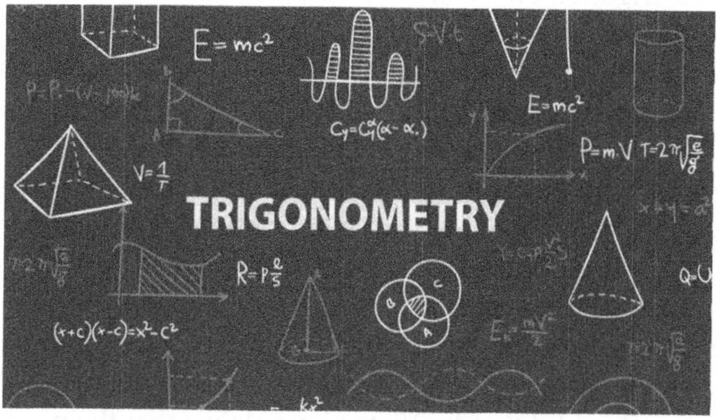

Objectives: This chapter

- *Introduces terminology associated with a right-angled triangle*
- *Explains the Pythagoras theorem for finding sides of a right-angled triangle*
- *Defines the trigonometric ratios sine, cosine, and tangent to find angles and the sides of a Right-angled triangle.*
- *Explains what is meant by trigonometric identities and gives a list of essential formulae*
- *Explains the concept of the Radian measure to find arc length and the area of a sector*
- *Introduces sine and cosine rules to find the trigonometric ratio to an angle greater than or less than 90°*
- *Explains bearing problems*

Ritu Shrivastava, Ramakant Bhardwaj and Satyendra Narayan. Mathematics for Engineers, (267–324)

7.1 Introduction

Tri (three) Gono (angels) Metry (To measure)

An important area of mathematics called trigonometry examines how triangles' sides and angles relate.

The word trigonometry comes from a 16th-century Latin derivative of the Greek words for the triangle (trigon) and measure (metron). In ancient times, the Greeks introduced trigonometry to answer questions like how far is it? And in what direction? Trigonometry deals with lines (which measure distance) and angles (which indicate direction).

Though the field emerged in Greece initially in the third century BC, the most critical developments and contributions (For example: the sine function) have come from India in the fifth century AD. It is unknown whether Indian scholars created trigonometry independently or under the influence of Greek scholars because the records of the earliest trigonometric discoveries made in Ancient Greece have been destroyed. A History of Mathematics, Third Edition by Victor Katz claims that the needs of Greek and Indian astronomers were the main driving forces behind the invention and development of trigonometry.

The lines and angles are formed into a triangle. Hence, the name "trigonometry" has become popular, meaning "triangle measurement" became famous. Only trigonometric methods are important to measure the object's height without scaling them.

For example: suppose you need to know the height of a university building without climbing it. It can be possible only by using some trigonometric concepts.

Trigonometric ideas are the foundation of the most technologically sophisticated engineering and physical science techniques. Right-angle triangles are the focus of trigonometry.

7.2 Pythagoras Theorem

$$(\text{hypotenuse})^2 = \text{opposite}^2 + \text{adjacent}^2$$

$$c^2 = a^2 + b^2$$

According to Greek mathematician Pythagoras, "the square of the hypotenuse equals the sum of the squares of the other two sides in any right-angle triangle."

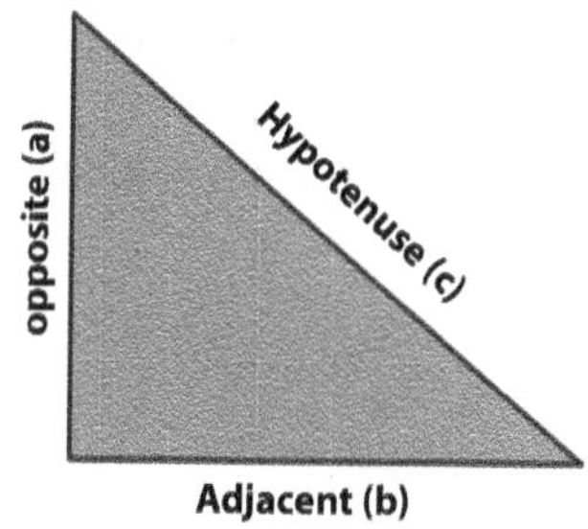

A right-angled triangle's third side can be calculated if we know the lengths of any two sides.

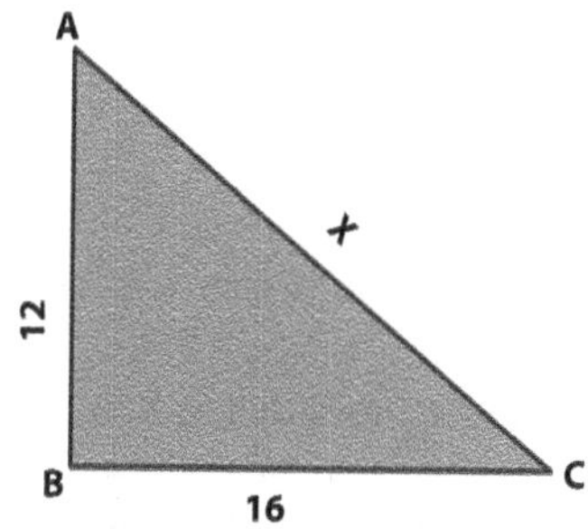

For example: to find the hypotenuse length in the given right-angle triangle.

let x be the length of the hypotenuse. By Pythagoras' Theorem.

$(\text{hypotenuse})^2 = \text{opposite}^2 + \text{adjacent}^2$

$(AC)^2 = (AB)^2 + (BC)^2$

$$x^2 = 12^2 + 16^2$$

$$x^2 = 144 + 256$$

$$x^2 = 400$$

$$x = \sqrt{400} = \pm 20$$

The hypotenuse is 20 units.

Interesting fact: *Pythagoras' theorem is a theorem that has more known proofs than any other theorem. The book Pythagoras Proposition by* ***Elisha Scroll Loomis*** *contains 367 proofs of it.*

Mixed level: Exercise 7.1 Practice Questions (Pythagoras Theorem)

1. Find the hypotenuse of a right-angled triangle when the other two sides are
   a. 5, 12
   b. 9, 12
   c. 35,12
   d. 7, 24
   e. 8, 15
   f. 9, 40
   g. 11,60
   h. 13, 84
   i. 33, 56
   j. 48, 55

2. Find the primitive Pythagorean triples up to 100. (Hint: (3,4,5), (5,12,13), etc.)
3. Show that if a triangle has sides of 8cm, 15cm, and 17cm, it is right-angled.
4. In a triangle, ABC, ∟B is a right angle, AB = 5.5cm, BC = 3.5cm. Find the diagonal AC.
5. A wall must be three feet away from a ladder's base. It is a 10-foot-long ladder. Does the ladder contact the wall above the ground?
6. A rectangular football field is 150 meters long and 130 meters wide. The coach instructs players to sprint diagonally from one corner to the other. How far do the players travel?
7. Sameer leaves the house to go to university. He walks 200m west and 125m north. How far away is he from his starting point? (Hint: calculate the length of the diagonal)
8. A walking track is in the shape of a rectangle. The dimensions of the track are 2km wide and 3km long. How much shorter is your walk if you walk diagonally across the track than along the two sides of it?
9. Aisha's house is 9km from Alia's and 40km from Fatima's. All three houses form a right-angle triangle at Sana's house. A road connects directly from Alia's house to Fatima's house. Find the length of the road.
10. The rectangular smartboard in a classroom is 157.5cm long. If the diagonal measures 200.5cm, how long is the width of the smartboard?

11. A runner runs 6km north and then 9km west. What is the shortest distance he must travel to return to their starting point? (Round your answer to the nearest whole).
12. To avoid the busy highway, Sameer must walk 6 km south and 10 km east. How many kilometers would be saved if it were possible to walk through the highway?
13. Calculate the diagonal length of a square whose sides are 42 meters.
14. Bassam travels 160 km south, then 36km east in his automobile. How far has he traveled since his beginning?
15. Firefighters have a 217-foot extension ladder. To reach 208 feet up a building, how far away from the building should the feet of the ladder be placed?

Think and answer.

16. The front door of Javed's room is 80 inches tall and 38 inches broad. He purchased a circular table that is 92 inches in diameter. The front door will the table fit through. Describe your response.
17. Sofia wants to check if the top of a side table she saw for sale is perfectly rectangular. She measured the dimensions of the table and found the following:
    Length of the table =18 inches, Width of the table = 9.5 inches. The diagonal length of the table = 19 inches. Is the table perfectly rectangular?
18. The rectangle ABCD represents the floor of a room (refer to the figure given). Ayesha stands at point P. Calculate the distance of Ayesha from
    a. Corner B of the room.
    b. The corner C of the room.

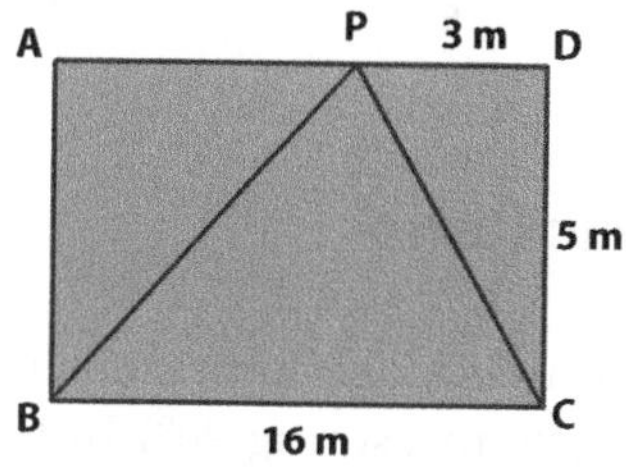

19. In the following diagram of a circle, O is the center, and the radius is 15cm. AB and CD are straight lines. Find the length of the CD if the length of OP is 3cm.

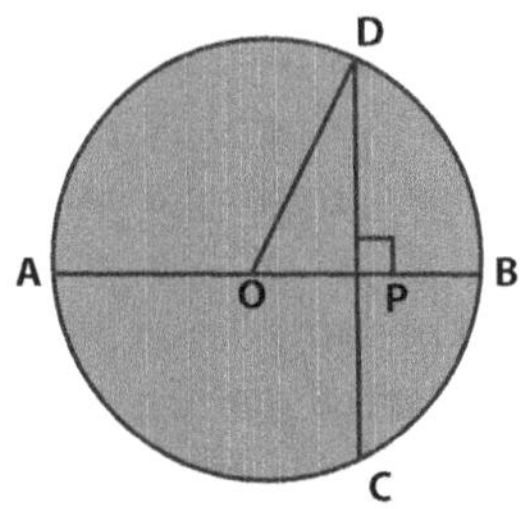

20. Two Bikers start from the same point. The First biker travels to the north and the second due to the east at the same speed. Find the distance of each biker from the starting point in kilometers per hour if, after 2 hours, they are 17km apart.
21. Shabnam wants to swim across a swimming pool that is 33 meters wide. She began swimming diagonally to the pool and covered 65 m. Find the length of the swimming pool.
22. The area of a square is 64 cm^2. Find the
    a. Length of a side.
    b. Length of the diagonal.
23. The two buildings are 36m and 45m tall, respectively. Find the distance between the tops of the two buildings if there is a 15m gap between them. (Hint: Look for AD in the given figure.)

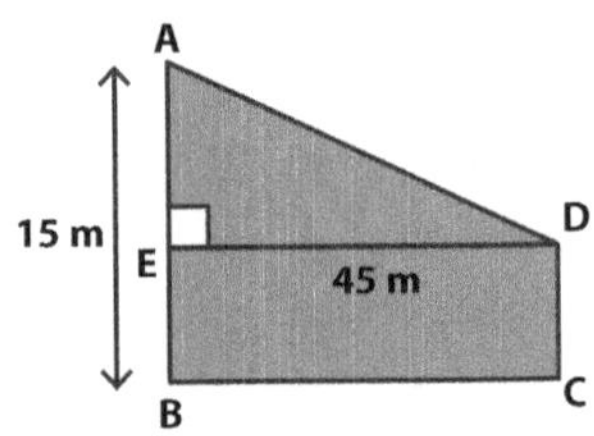

7.3 Trigonometric Ratio

The trigonometric ratios are unique measurements of a right triangle. (a triangle with one side angle measuring 90°) In this section, we'll look at different angle ratios for right triangles with sharp angles (an angle less than 90°). The third side of a right triangle (on the opposite side of the right angle) is known as the hypotenuse, and the two sides that make up the right angle are called the legs.

Consider a right triangle ABC where ∟B = 90° (right angle) and ∠A and ∠C are acute angles.

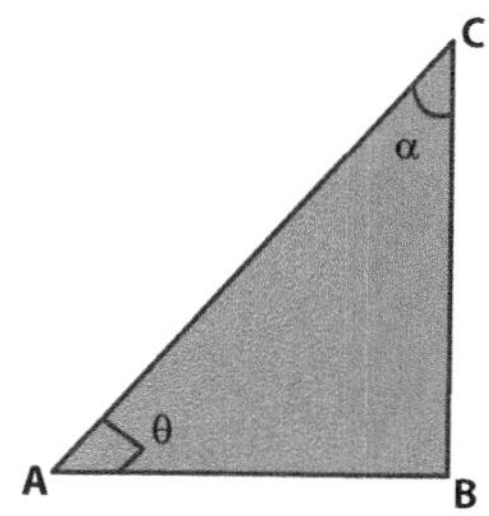

The following formulas are used to define the trigonometric ratios of angle A in the right triangle ABC:

$$\text{Sine A (sin A)} = \frac{\textit{Length of the leg opposite to angle A}}{\textit{Hypotenuse}} = \frac{BC}{AC}$$

$$\text{Cosine A (cos A)} = \frac{\textit{Length of the leg adjacent to angle A}}{\textit{Hypotenuse}} = \frac{AB}{AC}$$

$$\text{Tangent A (tan A)} = \frac{\textit{Length of the leg opposite to the angle A}}{\textit{Length of the leg adjacent to the angle}} = \frac{BC}{AB}$$

Trigonometric ratios are applicable only for a right triangle. There are three more trigonometric ratios.

$$\text{Cosecant A (cosec A)} = \frac{1}{\sin A} = \frac{\textit{Hypotenuse}}{\textit{Opposite}} = \frac{AC}{BC}$$

$$\text{Secant A (sec A)} = \frac{1}{\cos A} = \frac{\textit{Hypotenuse}}{\textit{Adjacent}} = \frac{AC}{AB}$$

$$\text{Cotangent A (cot A)} = \frac{1}{\tan A} = \frac{\textit{Adjacent}}{\textit{Opposite}} = \frac{AB}{BC}$$

Note that the ratios cosec A, sec A, and cot A are the reciprocal of the ratio sin A, cos A, and tan A. Also, observe that,

$$\text{Tan A} = \frac{BC}{AB} = \frac{\frac{BC}{AC}}{\frac{AB}{AC}} = \frac{sin\ A}{cos\ A}$$

So, we can say that $\quad \tan \text{A} = \dfrac{sin\ A}{cos\ A}$

Similarly, $\quad \cot A = \dfrac{cos\ A}{sin\ A}$

Note: *we usually use appreciated names of the function such as sin A, cos A, tan A, cosec A, sec A, and cot A. Symbol sin A is not the product of sin and 'A.' It is only the measurement of angle A; therefore, 'sin' separated from 'A' has no meaning. You may recall* ***SOH – CAH – TOA*** *to help you remember the functions.*

$$(SOH)\ sin A = \frac{0}{H} = \frac{Opposite\ side}{Hypotenuse}$$

$$(CAH)\ cos A = \frac{A}{H} = \frac{Adjacent\ side}{Hypotenuse}$$

$$(TOA)\ tan A = \frac{0}{A} = \frac{Opposite\ side}{Adjacent\ side}$$

The "Trigonometric Ratios Table" is given below. It helps us to find the value of standard trigonometric angles such as 0°,30°,45°,60°, and 90°. It consists of trigonometric ratios of sine, cosine, tangent, cosecant, secant, and cotangent.

7.4 Trigonometry Table

The trigonometric ratios of standard angles are essential to solving trigonometric problems. Therefore, it is necessary to remember the value of these ratios of these common angles. Also, you can use a calculator to get these values.

Trigonometry ratio table								
Angles in degree	0°	30°	45°	60°	90°	180°	270°	360°
Angles in radian	0	π/6	π/4	π/3	π/2	π	3π/2	2π
sin A	0	1/2	1/√2	√3/2	1	0	-1	0
cos A	1	√3/2	1/√2	1/2	0	-1	0	1
tan A	0	1/√3	1	√3	ND	0	ND	0
cot A	ND	√3	1	1/√3	0	ND	0	ND
sec A	1	2/√3	√2	2	ND	-1	ND	1
cosec A	ND	2	√2	2/√3	1	ND	-1	ND

*ND - not defined

Note: Trigonometric ratios of an angle will remain the same with the length of the sides of the triangles if the angle remains the same.

Interesting fact: *Trigonometry has a wide range of applications in other related scientific and mathematical disciplines. In the ancient era, trigonometry was carefully used and applied to estimate distances in space and on Earth.*

7.4.1 Angles

Recall the following definitions from elementary geometry.

(a) An angle is acute if it is between 0° and 90°.
(b) An angle is obtuse if it is between 90° and 180°.
(c) An angle is a right angle if it equals 90°.
(d) An angle is a straight angle if it equals 180°.

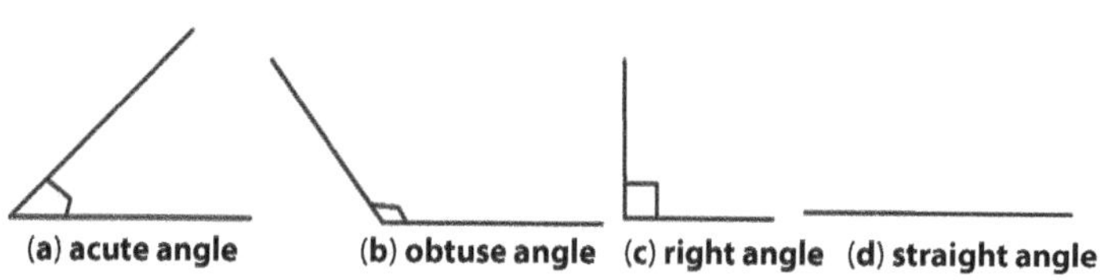

(e) In elementary geometry, we learned that the sum of the angles in a triangle equals 1800. ∠A+∠B+∠C=1800

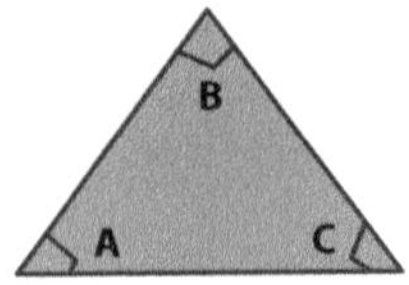

(e)

(f) The Iso-scalene triangle is a type of triangle where the lengths of the two sides are equal. AB = AC

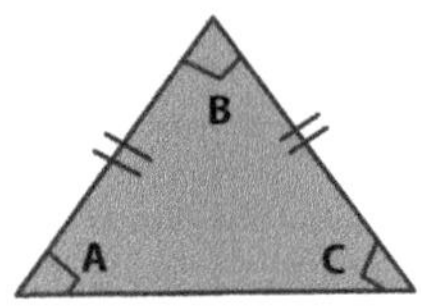

(f)

(g) In a right-angled triangle, one of the angles is 90°, and the other two angles are acute angles whose sum is 90°.

For example: to find the values of all the trigonometric ratios of angle Q

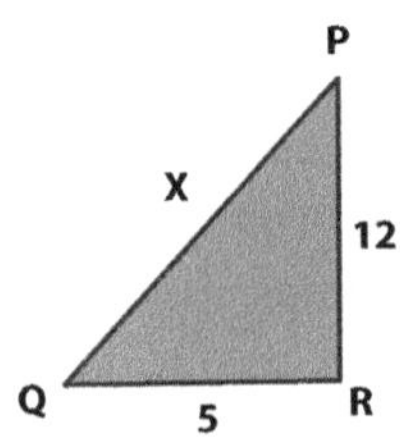

(g)

Before we can find the values of all six trig ratios, we need to find the length of the missing side.

$$x^2 = 12^2 + 5^2$$

$$x^2 = 144 + 25$$

$$x = \sqrt{169} = 13$$

Now, we can find the values of the six trig ratios, remember SOH, CAH, TOA

$$\sin\theta = \frac{0}{H} = \frac{12}{13} \qquad \text{cosec}\ \theta = \frac{H}{O} = \frac{13}{12}$$

$$\cos\theta = \frac{A}{H} = \frac{5}{13} \qquad \sec\ \theta = \frac{H}{A} = \frac{13}{5}$$

$$tan\theta = \frac{0}{A} = \frac{12}{5} \qquad \cot\ \theta = \frac{A}{O} = \frac{5}{12}$$

EXAMPLE 1: Use the triangle below to find all other trig ratios.

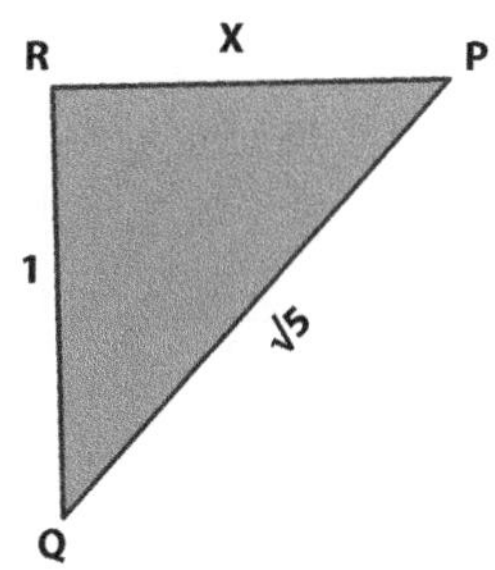

SOLUTION: To find the length of the missing side, use the Pythagoras theorem.

Observe that the length of the longest side is $\sqrt{5}$,. The side opposite Q is called the opposite, x, and the side adjacent to Q is 1.

Therefore, $(\sqrt{5})^2 = x^2 + (1)^2$

$$x^2 = (\sqrt{5})^2 - (1)^2$$
$$x^2 = 5 - 1$$
$$x^2 = 4$$
$$x = \sqrt{4} = 2$$

We can find the value of six trig ratios.

$$\sin\theta = \frac{0}{H} = \frac{2}{\sqrt{5}} \qquad \text{cosec } \theta = \frac{H}{O} = \frac{\sqrt{5}}{2}$$

$$\cos\theta = \frac{A}{H} = \frac{1}{\sqrt{5}} \qquad \sec\theta = \frac{H}{A} = \frac{\sqrt{5}}{1}$$

$$tan\theta = \frac{0}{A} = \frac{2}{1} \qquad \tan\theta = \frac{A}{O} = \frac{1}{2}$$

EXAMPLE 2: In the diagram below, find the value of sin θ, cos θ, and tan θ.
SOLUTION: Given that the Length of opposite = 5, Length of adjacent = 12, and Length of hypotenuse = 13

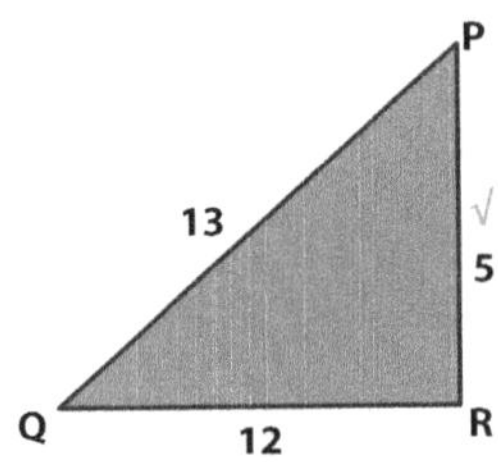

By using trig ratios – (SOH)

$$\sin\theta = \frac{Opposite\ side}{Hypotenuse} = \frac{5}{13}$$

By using trig ratio (CAH)

$$\cos\theta = \frac{Adjacent\ side}{Hypotenuse} = \frac{12}{13}$$

Similarly, tan θ (TOA)

$$\tan\theta = \frac{Opposite\ side}{Adjacent\ side} = \frac{5}{12}$$

EXAMPLE 3: Given that $\tan\theta = \frac{4}{3}$, find the trigonometric ratios sin and cos of the angle **θ**.
SOLUTION: Let's draw a right triangle PQR,

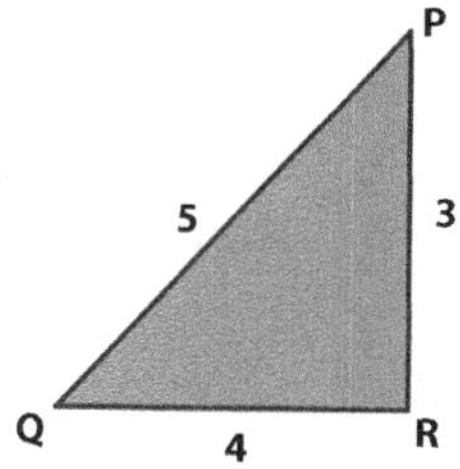

according to trig ratios

$$\tan \theta = \frac{opposite}{Adjacent} = \frac{AB}{BC} = \frac{4}{3}$$

Therefore, AB = 4, BC = 3
by using Pythagoras' theorem

$$PQ^2 = QR^2 + RP^2$$
$$PQ^2 = (4)^2 + (3)^2$$
$$PQ^2 = 16 + 9$$
$$PQ^2 = 25$$
$$PQ^2 = \sqrt{25} = 5$$

as we have all three legs of triangle PQR, we can write all the trig ratios using their properties.

$$\text{Sin A (SOH)} = \frac{0}{H} = \frac{4}{5}$$
$$\text{Cos A (CAH)} = \frac{A}{H} = \frac{3}{5}$$

EXAMPLE 4: Given $\sec A = \frac{7}{3}$, Find the other trig ratios of the angle θ.

SOLUTION: - Let's draw a right triangle PQR

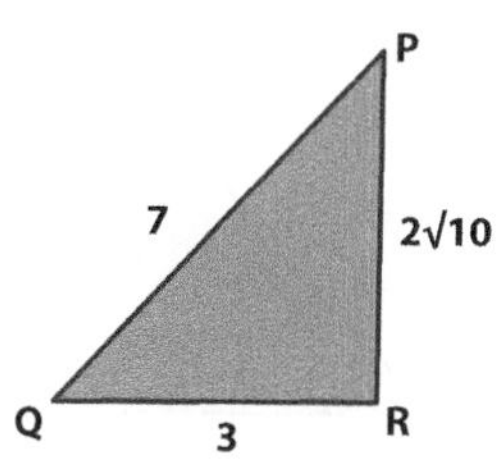

according to trig ratio properties

$$\sec A = \frac{Hypotenuse}{Adjacent} = \frac{7}{3}$$

Therefore, PQ = 7 and QR= 3
By using Pythagoras' theorem,

$$PQ^2 = QR^2 + RP^2$$
$$(7)^2 = (3)^2 + (RP)^2$$

$$(RP)^2 = (7)^2 - (3)^2 = 49 - 9 = 40$$

$$AB = \sqrt{40} = 2\sqrt{10}$$

As we have all three legs of triangle PQR, we can write all the trig ratios using their properties.

$$\text{Sin } \theta = \frac{O}{H} = \frac{2\sqrt{10}}{7} \qquad \text{cosec } \theta = \frac{H}{O} = \frac{7}{2\sqrt{10}}$$

$$\text{Cos } \theta = \frac{A}{H} = \frac{3}{7} \qquad \sec \theta = \frac{H}{A} = \frac{7}{3}$$

$$tan\theta = \frac{0}{A} = \frac{2\sqrt{10}}{3} \qquad \tan \theta = \frac{A}{O} = \frac{3}{2\sqrt{10}}$$

Mixed level: Exercise 7.2 Practice Questions (Trigonometric Ratio)

1. Find the values of all six trig ratios for angles A and C in the right-angle triangle ABC in the given figure.

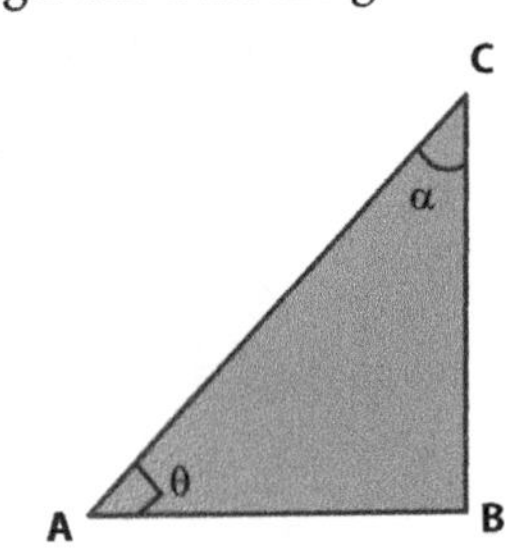

S. no.	Side AB	Side BC	Side AC	Angle θ	Angle α
a	5	12	13		
b	7	24	25		
c	8	15	17		
d	20	21	29		
e	9	40	41		
f	1	2	$\sqrt{5}$		
g	1	3	?		
h	2	5			

2. Find the value of the other trigonometric functions of the acute angle A given the indicated value of one of the functions.

   a. $\sin\theta = \frac{3}{4}$

   b. $\cos\theta = \frac{2}{3}$

   c. $\tan\theta = \frac{5}{9}$

   d. $\tan\theta = 3$

   e. $\cos\theta \frac{2}{\sqrt{10}}$

   f. $\sin\theta = \frac{1}{2}$

3. In the right triangle ABC, ∠ C = 440 and AC = 15cm. Find the length of each of the following. Round your answer to the nearest whole.
   a. AB
   b. BC

4. Use the trig ratios to find the figures' missing side lengths. Round your answer to the nearest whole.

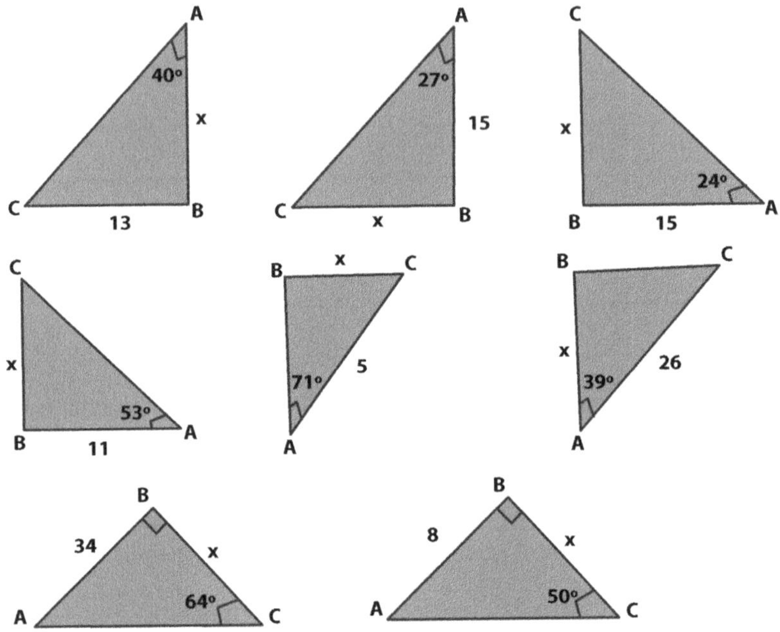

5. In the right triangle ABC, ∠A = 32° and BC = 24cm find the length of each of the following. Round your answer to one decimal place.
   a. AB.
   b. AC

Think and Answer

6. Can the sine of an angle ever equal 2? Why or why not?
7. An Isosceles triangle has legs of length 18cm and base angles that measure 46°. Find the height of the isosceles triangle.
(hint: create a right triangle by drawing the height)

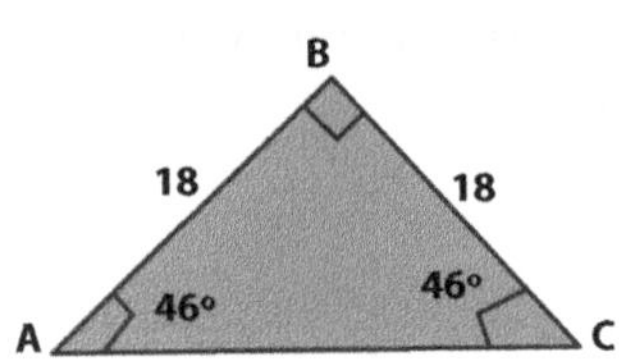

8. Javed walked 10km at an angle of 30° north of due east. How far East is Javed from his starting point?

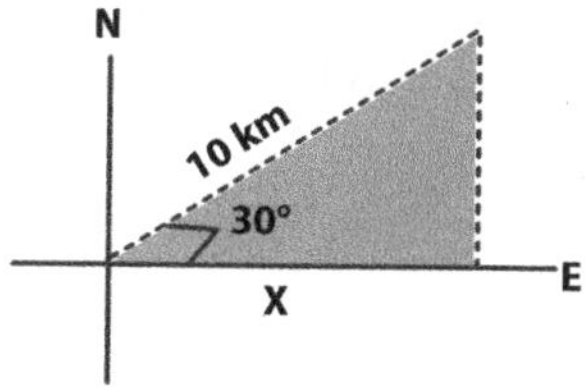

9. With the help of an Inclinometer, some students try to determine their height in their university building. They found that the angle of elevation is 54°. The horizontal distance of the building base from any fixed point on the ground is 40 feet. Find the height of the building.

Note: *An inclinometer is a tool used to measure the angles of a surface's slope, elevation, or depression.*

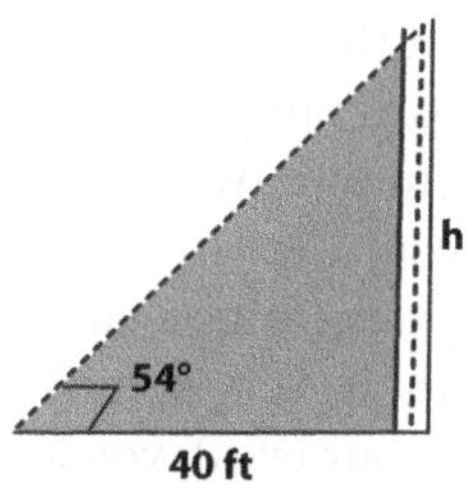

10. If the rays of the sun make a 34° angle with the ground, how long will the shadow of a tree be 20 feet high?

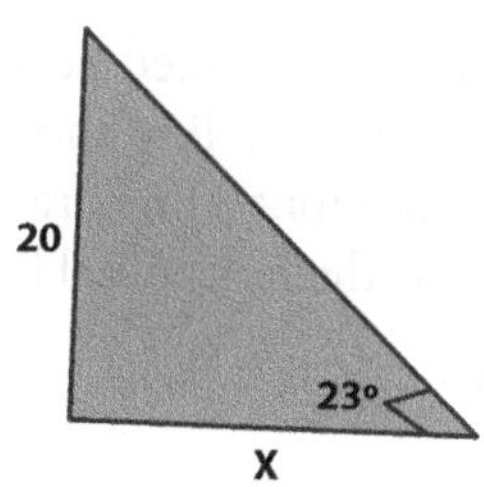

11. A person stands 250 ft away from a building and measures an angle of elevation of 33° from his horizontal line of sight to the top. Assume that the person's eyes are a vertical distance of 6 ft from the ground. What is the height of the building?

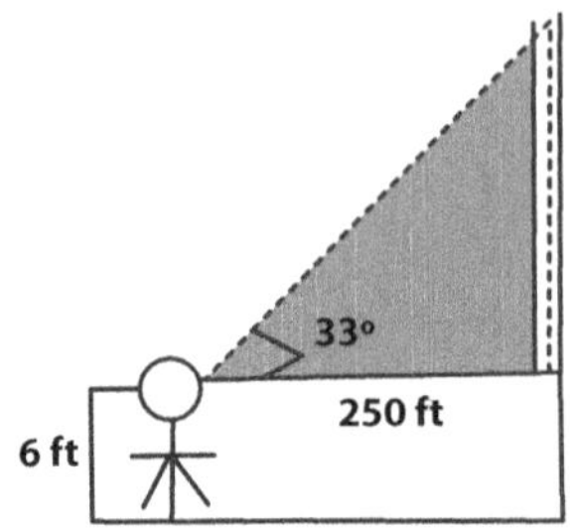

7.5 The Angle of Elevation and Angle of Depression

Right triangles utilize a wide range of designs in our everyday life. Examples are climbing stairs, looking at the rooftop, and crossing over a bridge; you see triangles in kites, skateboard ramps, road signs, airplanes, boats, mountains, and even a sandwich cut into two triangles, etc.

You "elevate" or "raise" your eyes in the sky whenever you look up at something. You are making something many call the "Angle of Elevation" with your eyes. Also, something very similar can be said when you look down (that is, you "depress" your eyes); you are making an edge called the "Angle of Depression."

7.5.1 Definition

The angle of elevation is the angle formed between the line of sight to an object above the observer's horizontal line and a horizontal line. The angle of depression is formed by a horizontal line drawn from the observer to the line of sight of an object below the horizontal line.

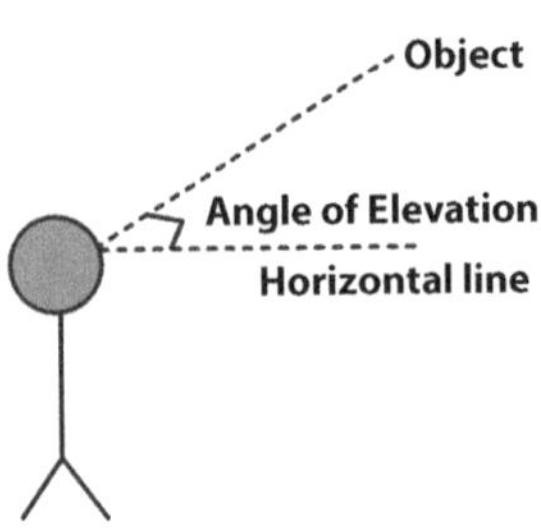

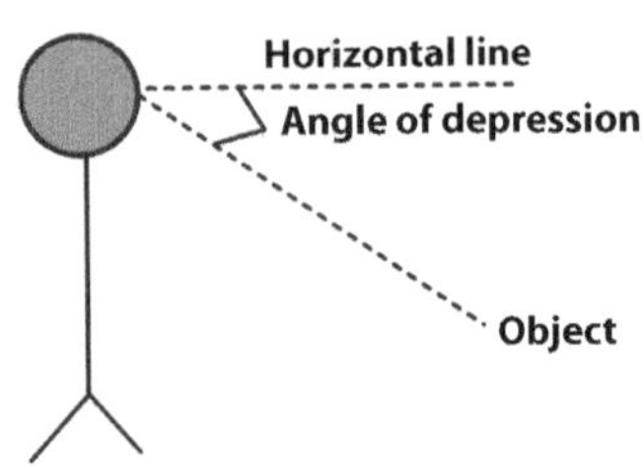

Practical example: Assume you have chosen to go camping with your friends. If you look at the hilltop from the camp on the ground, you are scared and ask your friend how far you need to travel to reach the top.

Your friend did not reply directly, but he hinted that the hill's height is 300 m, and the angle from your eyes to the top of the hill is 37°.

Is it possible for you to get the distance from your camp to the top of the hill?

yes, it is possible to calculate the distance from the camp to the top of the hill.

Let's understand the given scenario.

Let the distance from your eyes to the hilltop be x. Given that the elevation angle is 37° and the height of the hill = 300m. By using trigonometric ratios

$$\sin\theta = \frac{Opposite\ side}{Hypotenuse}$$

$$\sin 37° = \frac{300}{x}$$

$$x = \frac{300}{Sin\,37^0} = 500\text{ m}$$

Is this the correct distance that you calculated?

Check! You are missing something. What? Look at the picture again. It's given that the hill's height is 300 m from the ground, not from your eye level. So, you need to subtract the height from your eye level to the ground from the height of the hill. Let the height from your eye level to the ground be 1.8m.

Therefore, the length of the opposite = 300 – 1.8 = 298.2 m

$$\sin\theta = \frac{Opposite}{Hypotenuse}$$

$$\sin 37^{\circ} = \frac{298.2}{x}$$

$$x = \frac{298.2}{Sin 37^0}$$

$$x = 497\text{ m}$$

Think, what if your friend did not tell you the value of the angle of elevation? Will you be able to get it? Yes, with the help of a small device called an inclinometer.

EXAMPLE 5: The angle of depression of an object level with the base of the building from the top of the 40-meter-tall structure is 340 degrees. How far away is the object from the building's foundation?

SOLUTION: Let x m represent the separation between the object and the building's base. Let's draw a picture.

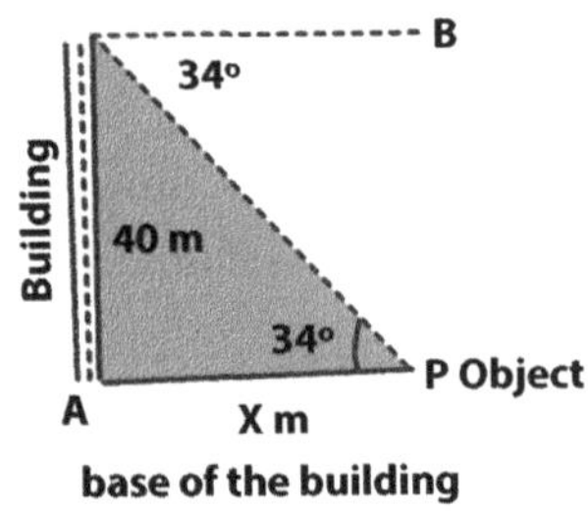

Given that the angle of depression = 34°

But $\angle APO = \angle BOP$

$\therefore \quad \angle APO = 34°$

from $\angle APO$, we have $\tan\theta = \dfrac{Opposite\,side}{Adjacent\,side}$

$$\tan 34° = \frac{40}{x}$$

$x \tan 34° = 40$ (by cross multiplication)

$$x = \frac{40}{tan\ 34^{\circ}}$$

$$x = 59.30\text{m (by using a calculator)}$$

The object is 59.30 m away from the base of the building.

EXAMPLE 6: Two buildings are 75 meters apart on the horizontal ground. The smaller structure is 14.5 meters tall. The top of the shorter building is 350 degrees lower than the top of the long building. Identify the taller building's height.

SOLUTION: let's draw a picture.

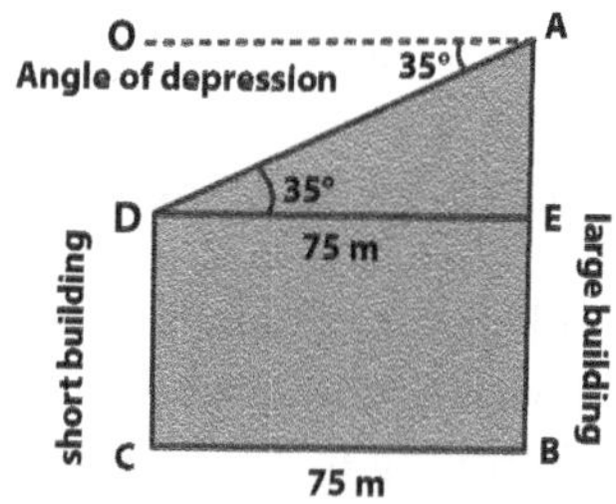

Step 1: draw a line from the top of the long building to the top of the shorter building. (line of sight)
Step 2: draw a horizontal line to the top of the long building and mark it at the angle of depression.
Step 3: Draw a horizontal line from the top of the shorter building to the long building.

Given that the angle of depression = 35°.
But ∠ OAD = ∠ ADE (Alternative angle)
∴ ∠ ADE = 35°
From ∠ AED, we have

$$\tan\theta = \frac{\textit{Opposite side}}{\textit{Adjacent side}} = \frac{x}{75}$$

$x = 75\times\tan 35°$
$x = 52.5$ m
And the remaining height of the long building is the same as the height of the shorter building, which is 14.5 m.
Hence the height of the longer building
= + 14.5
= 52.5 + 14.5 = 67m

Mixed level: Exercise 7.3 Practice Questions (The Angle of Elevation and Angle of Depression)

1. Ahmed is standing 5 m away from the tree, and he measured the angle of elevation to be 60°. Find out how tall the tree is. Assume that Ahmed's height is 1.7 m up to his eyes.

2. Aalia is swimming in the pool and notices an object underwater below her. The angle of depression is 38°, and the depth of the pool at that point is 3.5m. How far is she from the object?
3. There is a 65° angle of depression from the tree's top to a car's base. How far away is the car if the tree is 70 feet tall from the tree base?
4. A historical tower leans at a 5° angle and has a vertical height of 60m. How tall was the tower when it was built?
5. A tree struck by lightning in a storm break and falls over to form a triangle with the ground. The tip of the tree makes a 41° angle with the ground, and it is 28 feet from the base of the tree. What was the height of the tree?
6. Isa is flying his kite and notices that he has a lot out of the entire 100 feet of string. The angle his string makes with the ground is 45°. How high is his kite at this time?

7.6 Trigonometric Equation

7.6.1 Definition

Equations that involve trigonometric functions are known as trigonometric equations. Before understanding trig equations, we need to know a few definitions.

7.6.2 Oriented Angle

In an oriented angle, the direction of the rotation is essential. See the figure below if the angle from the initial side is swept out with the angle at a terminal side.

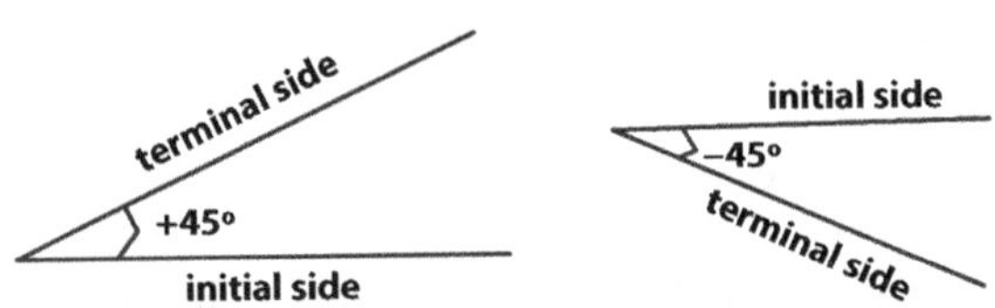

When the angle is anti-clockwise, we say that the angle is "positive." When the rotation is clockwise, we say the angle is "negative."

7.6.3 Identities

Several relations between trigonometric functions are valid for all values of the given angles. Such relations are said to be identically true or to be Identities. Some negative angle identities are well-defined and used as a formula. These identities are shown below.

Negative Angle Identities

$\sin(-x) = -\sin x$
$\cos(-x) = \cos x$
$\tan(-x) = -\tan x$
$\operatorname{cosec}(-x) = \operatorname{cosec} x$
$\sec(-x) = \sec x$
$\cot(-x) = -\cot x$

These identities show that sin θ and tangent are odd functions as $f(-x) = -f(x)$. And the cosine is an even function as $f(-x) = f(x)$. Trigonometric identities play an important role in Calculus. Any function which displays exact preparation, or a fixed pattern is called a periodic function. For both sin θ and cos θ, the period is 360°, while for tan θ, the period is 180°.

There are a few more trigonometric identities that are infrequently used. Here is a list of important trigonometric identities that will help us quickly recognize the situation where they will be useful while working on a problem.

Ratio Identities

$$\operatorname{cosec}\theta = \frac{1}{\sin\theta}$$

$$\sec\theta = \frac{1}{\cos\theta}$$

$$\tan\theta = \frac{\sin\theta}{\cos\theta}$$

$$\cot\theta = \frac{\cos\theta}{\sin\theta}$$

$$\cot\theta = \frac{1}{\tan\theta}$$

Pythagoras Identities

(1) $\sin^2\theta + \cos^2\theta = 1$
$\sin^2\theta = 1 - \cos^2\theta$
$\cos^2\theta = 1 - \sin^2\theta$

(2) $1 + \tan^2\theta = \sec^2\theta$
$\tan^2\theta = \sec^2\theta - 1$
$\sec^2\theta - \tan^2\theta = 1$

(3) $\cot^2\theta = \text{cosec}^2\theta - 1$
$\text{cosec}^2\theta - \cot^2\theta = 1$
$\text{cosec}^2\theta = \cot^2\theta + 1$

Complementary/Supplementary Angle Identities

$\sin\theta = \cos(90° - \theta)$
$\tan\theta = \cot(90° - \theta)$
$\sec\theta = \text{cosec}(90° - \theta)$
$\sin\theta = \sin(180° - \theta)$

Compound Angle Identities

$$\sin(A+B) = \sin A\cos B + \cos A\sin B$$

$$\sin(A - B) = \sin A\cos B - \cos A\sin B$$

$$\cos(A+B) = \cos A\cos B - \sin A\sin B$$

$$\cos(A - B) = \cos A\cos B + \sin A\sin B$$

$$\tan(A+B) = \frac{\tan A + \tan B}{1 - \tan A \tan B}$$

$$\tan(A-B) = \frac{\tan A - \tan B}{1 + \tan A \tan B}$$

Double Angle Identities

$$\sin 2A = 2\sin A\cos A$$

$$\cos 2A = \cos^2 A - \sin^2 A = 1 - 2\sin^2 A = 2\sin^2 A - 1$$

$$\tan 2A = \frac{2\tan A}{1 + \tan^2 A}$$

Half Angle Identities

$$\sin\frac{A}{2}=\sqrt{\frac{1-\cos A}{2}}$$

$$\cos\frac{A}{2}=\sqrt{\frac{1+\cos A}{2}}$$

$$\tan\frac{A}{2}=\frac{\sin A}{1+\cos A}$$

$$=\frac{1-\cos A}{\sin A}$$

7.6.4 Quadrants

As shown in the picture below, the coordinate axes divide the plane into four quadrants: First, Second, Third, and Fourth.

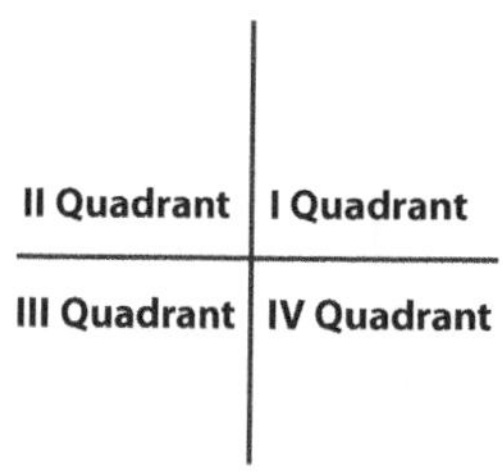

Each point that is located in one of the four quadrants can have its trigonometric ratio's sign determined by looking at its X and Y coordinates.

> *Remember,*
> *Angles between 0° and 90° are in the I quadrant.*
> *Angles between 90° and 180° are in the II quadrant.*
> *Angles between 180° and 270° are in the III quadrant.*
> *Angles between 270° and 360° are in the IV quadrant.*

These are summarized in the diagram below.

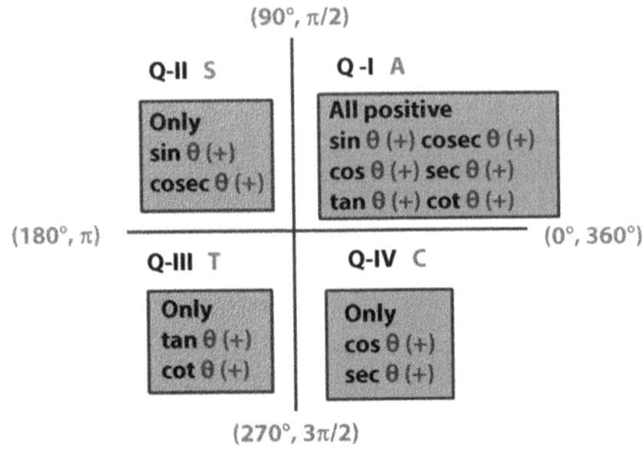

You can easily understand the same by the table below.

Quadrants	sin A	cos A	tan A
I	+	+	+
II	+	-	-
III	-	-	+
IV	-	+	-

Interesting fact: *The all sin, cos, and tan rule can be remembered quickly using the following phrases.*

All Silver Tea Cup Or All Students Take Calculus

EXAMPLE 7: Complete the following table with positive and negative signs without using a calculator.

θ	sin A	cos A	tan A
250°			
150°			
112°			
72°			
85°			
287°			
310°			
185°			
98°			

SOLUTION:

θ	sin A	cos A	tan A
250°	-	-	+
150°	+	-	-
112°	+	-	-
72°	+	+	+
85°	+	+	+
287°	-	+	-
310°	-	+	-
185°	-	-	+
98°	+	-	-

EXAMPLE 8: Write the following in terms of an acute angle.

(a) Cos 120°

SOLUTION: Since $\theta = 120°$ is more than 90°

So that it lies in the II quadrant.

Since the angle in the II quadrant is $180 - \theta$

$\cos(120°) = \cos(180° - \theta)$

$= \cos(180° - 120°)$

$= \cos 60°$

But in the II quadrant cos will be negative

Hence $\cos 120° = -\cos 60°$

(b) Sin (335°)

SOLUTION: since $\boldsymbol{\theta} = \mathbf{335°}$ is more than 270°

Hence, the angle is lying in the IV quadrant.

The value of the angle in the IV quadrant is.

$360° - \theta$

$= 360° - 335°$

$= 25°$

But the value of sin in the IV quadrant is negative

Therefore, $\sin(335°) = -\sin(25°)$

EXAMPLE 9: Solve $2 \sin x \cos x = \cos x$ in the interval 0° to 360°

SOLUTION: $2 \sin x \cos x = \cos x$

Step 1: Divide Cos x both Sides

$$\frac{2 \sin x \cos x}{\cos x} = \frac{\cos x}{\cos x}$$

$$2 \sin x = 1$$

$$\sin x = \frac{1}{2}$$

$$x = \sin^{-1}\left(\frac{1}{2}\right)$$

$$x = 30°$$

EXAMPLE 10: Find all solutions to the equation.
SOLUTION: $2\sin^2 x - \sin x - 1 = 0$ in the interval $[0, 2\pi]$.

$2\sin^2 x - \sin x - 1 = 0$

Let $\sin x = x$, and convert the trigonometric equation into the quadratic equation.

$$2x^2 - x - 1 = 0$$
$$2x^2 - 2x + x - 1 = 0$$
$$2x(x - 1) + 1(x - 1) = 0$$
$$(2x + 1)(x - 1) = 0$$

Now replace x with $\sin x$

$$2\sin x + 1 = 0 \quad \text{or} \quad \sin x - 1 = 0$$

$$\text{Sin } x = \frac{-1}{2} \qquad \text{or} \quad \sin x = 1$$

$$x = \sin^{-1}\left(-\frac{1}{2}\right) \text{ or } x = \sin^{-1}(1)$$

$$x = \pi/6 \qquad \text{or } x = \pi/2$$

sin is negative in the III and IV quadrants

Q – III	Q –IV	Q - II
$\pi + \frac{\pi}{6}$	$2\pi - \frac{\pi}{6}$	$\pi - \frac{\pi}{2}$
$\frac{7\pi}{6}$	$\frac{11\pi}{6}$	$\frac{\pi}{2}$

How to solve trigonometric equations? Have a look at the quadrant diagram.

Q - II	Q – IV
$180° + \theta$	$360 - \theta$
$180° + 61 = 241°$	$360 - 61 = 299°$

II Quadrant **(180° - θ)**	**I Quadrant** **All positive (X = θ)**
III Quadrant **(180° + θ)**	**IV Quadrant** **(360° - θ)**

To solve the trig equation into interval $0° \leq x \leq 360°$. First, you need to find the angle θ in the I quadrant by taking its absolute value; later, you choose the quadrant in which the trig function has the correct sign.

EXAMPLE 11: for interval $0° \leq x \leq 360°$ find the other values of θ, when $\tan\theta = -1.8$

SOLUTION: Step 1: Find the angle θ in the I quadrant.

$\tan\theta = -1.8$

$\theta = \tan^{-1}(-1.8) = 60.94 = 61°$ (to the nearest degrees)

Since tangent is negative in the II and IV quadrants

The solutions are $\theta = 241°$ or $299°$

EXAMPLE 12: For interval $-180° \leq x \leq 180°$, Find other values of θ when, $\cos\theta = 0.61$

SOLUTION:

Step: 1 Find the angle in the first quadrant

$$\cos\theta = 0.61$$

$\theta = \cos^{-1}(0.61)$

$\theta = 52°$ (To the nearest degree)

Since, cosine is +ve in IV quadrant

So, we get the value of the angle in the IV quadrant

$$= 360° - \theta$$

$= 360° – 52$
$= 308°$
The solution $\theta = 308°$ is in the IV quadrant and not in the interval $-180 \leq \theta \leq 180$, so we subtract 360° from the obtained value of θ.
$\theta = 308° – 360° = -52°$

Note: If you are asked to find solutions in $0 \leq x \leq 2\pi$, then you must convert the answers to radians.

Mixed level: Exercise 7.4 Practice Questions (Trigonometric Equations)
Find the value of x in the given interval

1. $2 \sin^2 x – \sin x – 1 = 0$ $(0° \leq x \leq 360°)$
2. $\sin x = 0.82$ $(0 \leq x \leq 360°)$
3. $\tan x = -0.94$ $(0 \leq x \leq 360°)$
4. $\cos x = -0.49$ $(0 \leq x \leq 360°)$
5. $\sin x = -0.23$ $(0 \leq x \leq 360°)$
6. $\cos x = 0.77$ $(0 \leq x \leq 2\pi)$
7. $\sin x = 0.61$ $(0° \leq x \leq 360°)$
8. $\cos x = 0.39$ $(-180° \leq x \leq 180°)$
9. $\sin x = -0.71$ $(0 \leq x \leq 2\pi)$
10. $\cos x = -0.68$ $(0 \leq x \leq 360°)$

7.7 Radian Measure

In lower class, we usually learn to measure an angle in degrees. However, there are other ways of measuring an angle called radians. In many scientific and engineering calculations, radians are used in preference to degrees. We understand a full rotation of an angle of a circle is 360°.

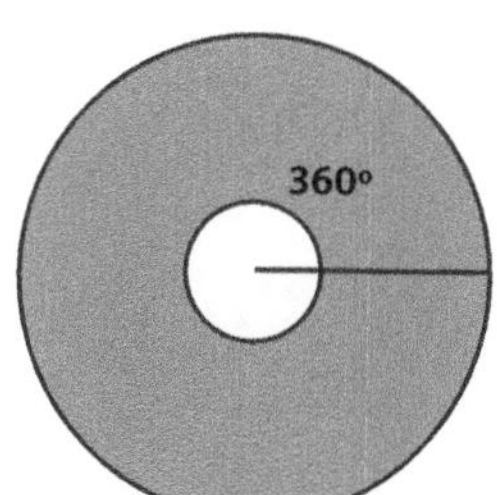

To measure an angle in radians, Consider a circle of radius r. In the figure, the arc shown has a length chosen to equal that radius.

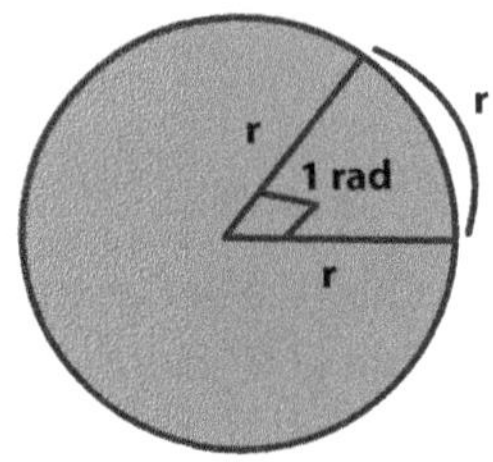

The angle is 1 radian.

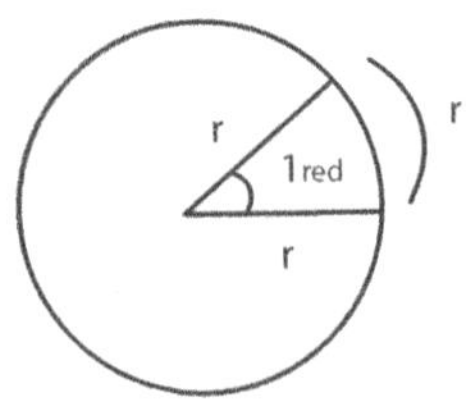

An arc of length r, as shown in the picture, subtends an angle of 1 radian; hence, an arc of length 2r will do the same for an angle of 2 radians.

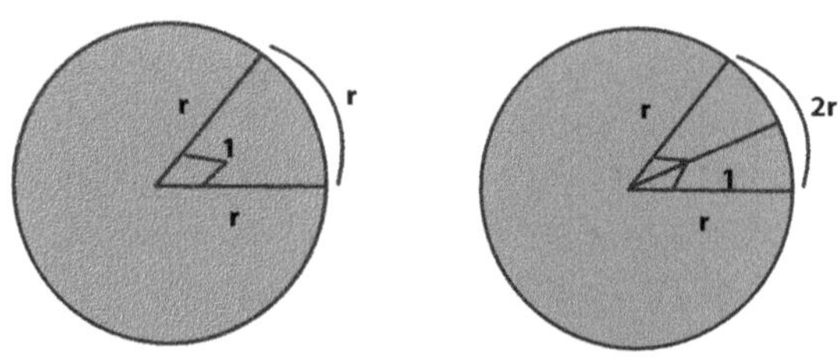

In the general case referring to the figure below, the length l of an arbitrary arc that subtends an angle θ is $r\theta$.

They can be expressed as $l = r\theta$

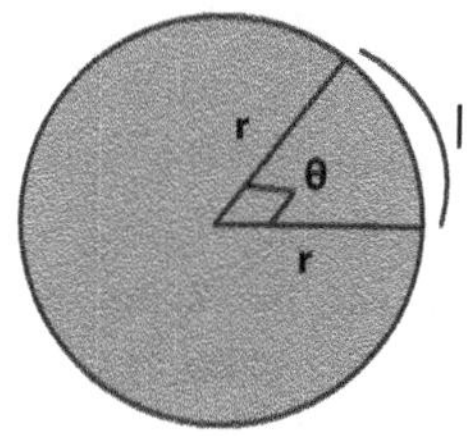

This relationship is considered a formula to calculate the length of the arc. Arc length $l = r\,\theta_{rad}$

Degree to Radian Multiply the given value by $\frac{\pi}{180^\circ}$	Radian to Degree Multiply the given value by $\frac{180^\circ}{\pi}$

Where r denotes the radius of a circle and θ denotes the angle in radians
Note: *When directions are stated in terms of degrees, they are typically found by starting at "North," which is "0°," and traveling dock-wise by the specified number of degrees. N 36° W or S 27° E are two other degree-based ways to give directions. They mean 36 degrees west of north and 27 degrees east of south.*

Degrees are technically not numbers. For any mathematical calculation, we need numbers. The 360° for one revolution is the same as 2π in radians. Because the circumference C of a circle with radius r is

$C = 2\pi r$ If $r = 1$, $C = 2\pi$. The numerical value of 360° is 2π.

7.7.1 Converting Between Radians and Degrees

We use radians and degrees for different aspects; both are important in mathematics.

For example: if you describe the direction, you will use degrees like turn 60° to reach that point. But if you need to find a sector's area or a circle's arc length, you will need the same numerical value, Radian. You have that value in Radians that you can use directly in the formula to calculate your answer. Some problems can only be solved by using radians, so what will you do if the values are given in radians?

You need to do the conversion from degrees to radians.

Here is an easy way to do so.

Note: *Radians and degrees are two types of units for measuring angles. There are many more such units, like gradients and MRAD, but degrees and radians are the ones we mostly use.*

Note: *Degrees are used to express both direction and the side of the angle. If you stand facing directly to the north, you are facing the direction to zero degrees, written as 0° (the superscripted "circle" stands for "degree"). If you turn yourself entirely around, you end up facing north again; we have turned through 360°.* That is, one complete revolution (or one circle) is 360°.

A complete revolution is 360°, and a half turn (or an about-face) is 180°. Look at these figures.

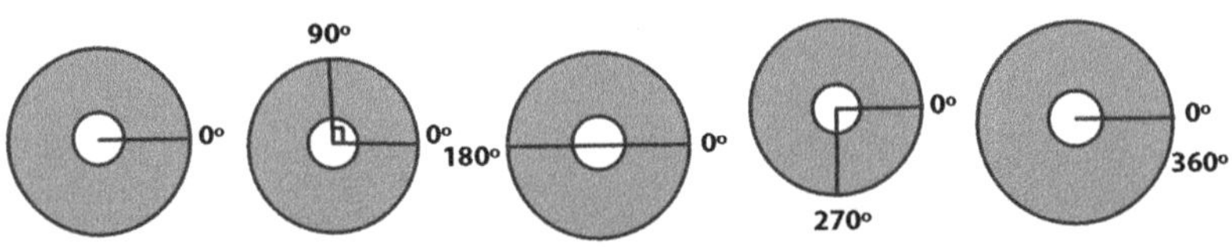

EXAMPLE 13: convert 270° into radians.

SOLUTION: To convert degrees to Radians, we need to multiply by $\frac{\pi}{180^\circ}$

Therefore, $\frac{270^\circ}{1} \times \frac{\pi}{180^\circ} = \frac{3\pi}{2}$

Hence, the radian measure of 270° is $\frac{3\pi}{2}$.

EXAMPLE 14: convert $\frac{\pi}{6}$ into degrees.

SOLUTION: To convert Radian to Degree, we need to multiply by $\frac{180^\circ}{\pi}$

Therefore, $\frac{\pi}{6} \times \frac{180^\circ}{\pi} = \frac{\pi}{\pi} \times \frac{180^\circ}{6} = 1 \times \frac{180^\circ}{6} = 30^\circ$

Hence, the degree measure of $\frac{\pi}{6}$ is 30°.

EXAMPLE 15: convert $\frac{11\pi}{3}$ to degree

SOLUTION: To convert Radian to Degree, we need to multiply by $\frac{180^\circ}{\pi}$

Therefore $\frac{11\pi}{3} \times \frac{180^\circ}{\pi}$

$$= \frac{11}{3} \times \frac{180^\circ}{1} \times \frac{\pi}{\pi}$$

$$= \frac{11 \times 180^\circ}{3} \times 1 = 600^\circ$$

Hence the degree measure of $\frac{11\pi}{3}$ is 660°.

EXAMPLE 16: convert 120° into Radian.
SOLUTION: To convert Degrees into Radians, we need to multiply by $\frac{\pi}{180^\circ}$

Therefore, $120^\circ \times \frac{\pi}{180^\circ}$

$$= \frac{120^\circ}{180^0} \times \pi = \frac{2}{3}\pi$$

Hence, the radian measure of 120° is $\frac{2}{3}\pi$.

7.7.2 Area of Sector

Sector: An area of a circle that is bounded by two radii and the connecting arc is known as a sector. Consider a pizza slice that is a sector. All sectors have a central angle. This is the angle that subtends to the center of the circle. The sector area can be found in a couple of different ways depending on what you know; you will always need to know the radius.

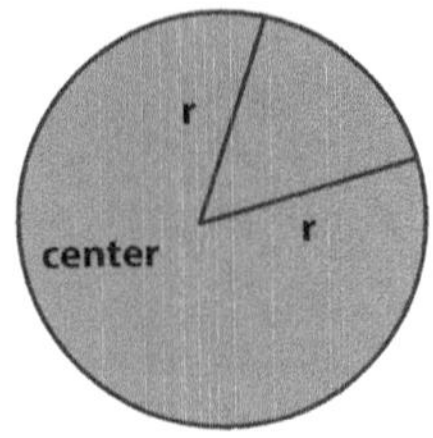

Major Sectors and Minor Sectors

A **major** sector is a sector
with a **bigger area.**
A **minor** sector is a sector with a **smaller area.**

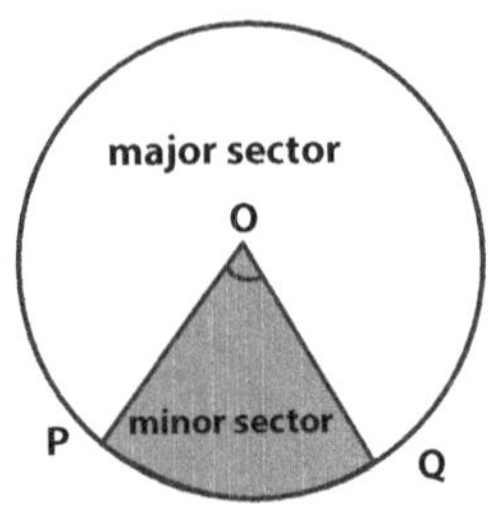

We know that
The area of the entire circle = $\pi\, r^2$

And the area of the half-circle $= \frac{1}{2}\pi\, r^2$

The principle of the area of the sector follows this same logic.

So, if you know the central angle, the area will be $\frac{n}{360} \times \pi r^2$ (in degree)

Where n is the number of degrees in the central angle, so if $n = \theta$,

$$A = \frac{\theta}{2\pi} \times \pi r^2 \text{ (in Radian)}$$

We can use the direct formula for calculating

Area of sector $= \frac{1}{2} r^2\, \theta_{\text{rad}}$

EXAMPLE 17: Given that the circle radius is 5cm, calculate the area of the sector with an angle of 60°.
SOLUTION: we know that

$$\text{Area of the sector}_{\text{(in Radian)}} = \frac{1}{2} r^2 \, \theta_{rad}$$

$$\text{And the area of the sector}_{\text{(in degree)}} = \frac{\theta}{360^0} \times \pi r^2$$

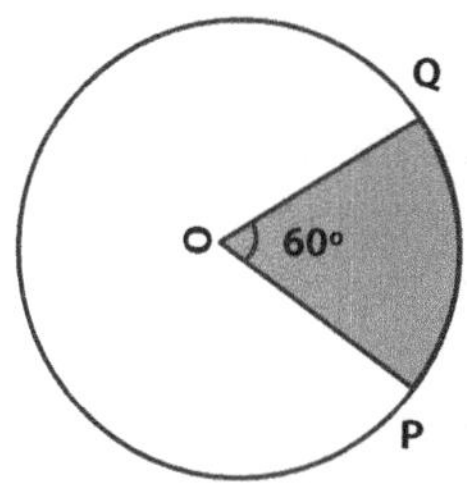

Since the angle is given in degrees, we can use the second formula

$$= \frac{60°}{360°} \times \pi (5)^2$$

$$= \frac{2.5}{6} \pi$$

$$= 13.09 \text{ cm}^2$$

Another way we can use the formula is when θ is given Radian, or first, we can convert θ into Radians if given in degrees.

$$\theta = 60°$$

$$\theta_{\text{rad}} = 60° \times \frac{\pi}{180°}$$

$$\theta_{\text{rad}} = \frac{1\pi}{3}$$

$$\text{Area of sector} = = \frac{1}{2} r^2 \, \theta_{\text{rad}}$$

$$= \frac{1}{2}(5)^2 \times \frac{\pi}{3}$$

$$= \frac{25\pi}{6} = 13.09 \text{ cm}^2$$

EXAMPLE 18: The area of the sector is 45.5 cm². When the radius of a circle is 5 cm, calculate the angle of the sector.

SOLUTION: we know that

$$\text{The area of sector} = \frac{\theta}{360^\circ} \times \pi r^2$$

$$45.5 = \frac{\theta}{360^\circ} \times \pi(5)^2$$

$$\theta = \frac{45.5}{25\,\pi} \times 360^\circ = 208.5^\circ$$

Or

$$\text{Area of sector} = \frac{1}{2} \pi r^2 \, \theta_{\text{rad}}$$

$$45.5 = \frac{1}{2} \times \pi(5)^2 \, \theta_{\text{rad}}$$

$$\theta_{\text{rad}} = \frac{45.5 \times 2}{25}$$

$$\theta_{\text{rad}} = 3.64$$

$$\theta^\circ = 3.64 \times \frac{180^\circ}{\pi} = 208.5$$

Mixed level: Exercises 7.5 Practice Questions (Radian Measure)

1. If the angle formed by two radii is 108° and the radius of a circle is 7 cm works out the area of the sector and arc length.
2. From the below figure (not on scale), find the area of sector POQ. round your answer up to 2d.p.

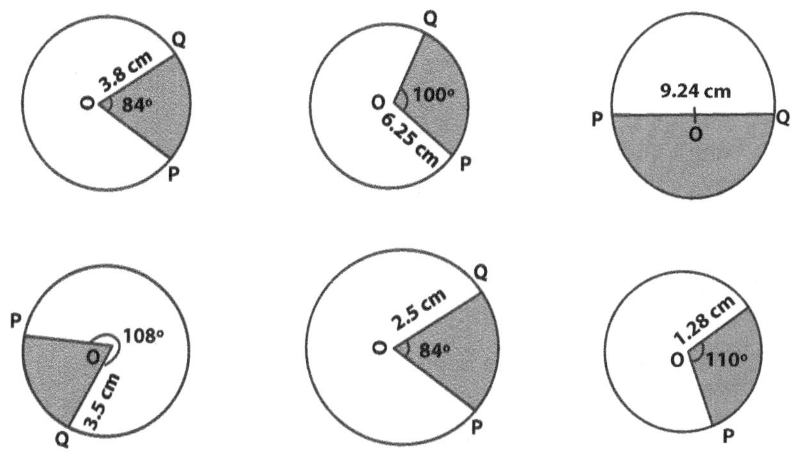

3. From the figures below(not on scale), find the missing angle if the area of the sector is 225 cm².

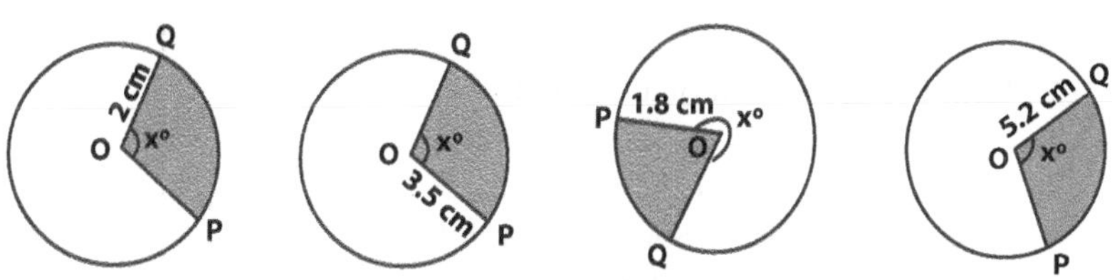

4. Find the lengths of each arc given below in the figure. Round your answer to the nearest whole.

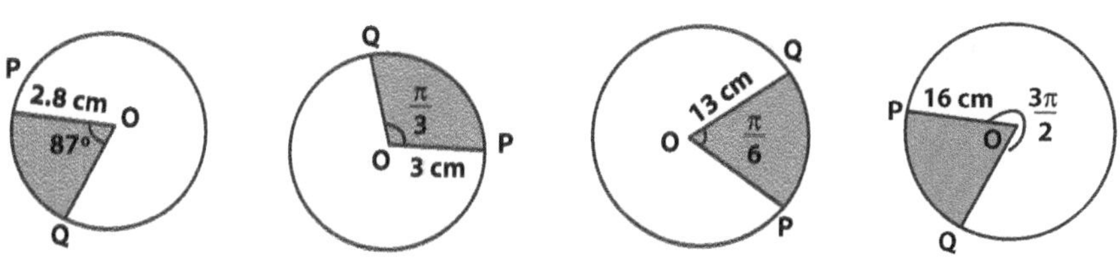

5. Calculate the arc's length and a sector's area for the cases below.
   a. $r = 18cm, \theta = 60°$.
   b. $r = 16cm, \theta = 75°$.
   c. $r = 9cm, \theta = \frac{7\pi}{4}$
   d. $r = 14cm, \theta = \frac{19\pi}{12}$
   e. $r = 10cm, \theta = \frac{\pi}{2}$
   f. $r = 12cm, \theta = \frac{5\pi}{3}$
   g. $r = 7km, \theta = 60°$

6. Find the area of the sector if
   a. Arc length = 16cm and $\theta = 64°$
   b. Arc length = 20cm and $\theta = \frac{\pi}{3}$
   c. Arc length = 45cm and $\theta = \frac{2\pi}{3}$
   d. Arc length = 108cm and $\theta = 70°$
   e. Arc length = 94cm and $\theta = 108°$
   f. Arc length = 85cm and $\theta = 45°$

7. Find the arc length of a circle if.
   a. Area of sector = $225cm^2$ and r = 21 cm
   b. Area of sector = $125cm^2$ and r = 9 cm
   c. Area of sector = $185cm^2$ and r = 11 cm
   d. Area of sector = $504cm^2$ and r = 16 cm
   e. Area of sector = $110cm^2$ and $\theta = 35°$
   f. Area of sector = 306 cm^2 and $\theta = \frac{5\pi}{2}$
   g. Area of sector = 306 cm^2 and $\theta = \frac{\pi}{4}$

Think and Answer:

8. A circular fountain of a 3cm radius lies alone in the center of a circular park of a 100cm radius. Calculate the total walking area available to the visitors visiting the park.
9. The arm and blade of the windscreen wiper are 28 inches long. If the blade is 21 inches long and the wiper sweeps out an angle of 110°, how much window area can the blade clean?
10. For a circle with radius r, the arc length corresponding to the central angle measuring $x°$ is L, and the area of the related sector is A. In terms of initial values, what would be the arc length and the sector area for a circle with a radius of 2r and the same central angle?

7.8 The Sine and Cosine Rules

Pythagoras' theorem allows us to handle any given right triangle problem easily. But what if the triangle isn't a right triangle? In some instances, we can use the laws of sine and cosine.

7.8.1 The Sine Rule

$$\frac{\sin A}{a} = \frac{\sin B}{b} = \frac{\sin C}{c} \quad Or \quad \frac{a}{\sin A} = \frac{b}{\sin B} = \frac{c}{\sin C}$$

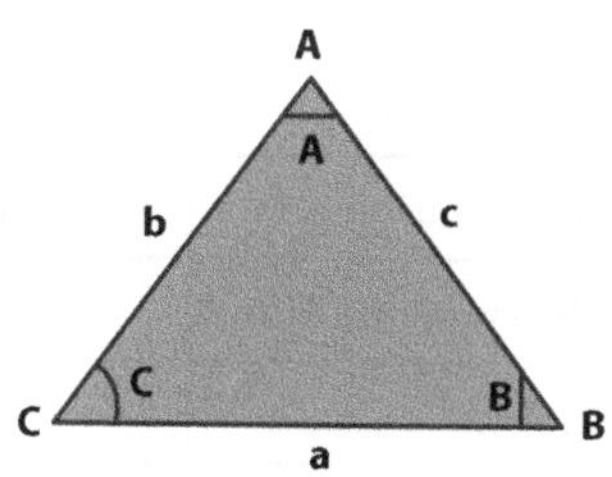

A triangle's angles are A, B, and C, while its sides' lengths are a, b, and c. Sine law helps us to solve triangles in the angle-angle-side (AAS), the angle-side-angle (ASA), and the angle—side—side (ASS) cases.

Cosine rules help solve other cases like Side–Angle–Side (SAS) and Side- Side- Side (SSS).

7.8.2 The Cosine Rule

(To calculate sides)

$a^2 = b^2 + c^2 - 2bc \cos A$

$b^2 = a^2 + c^2 - 2ac \cos B$

$c^2 = a^2 + b^2 - 2ab \cos C$

(To calculate angles)

$$\cos A = \frac{b^2 + c^2 - a^2}{2bc}$$

$$\cos B = \frac{a^2 + c^2 - b^2}{2ac}$$

$$\cos C = \frac{a^2 + b^2 - c^2}{2ab}$$

$0° \leq A \leq 90°$	**$90° \leq A \leq 180°$**
$a \leq b \sin A$: No solution	$a \leq b$: No solution
$a = b \sin A$; One solution	$a > b$: one solution
$b \sin A \leq a \leq b$; Two solution	
$a \geq b$: one solution	

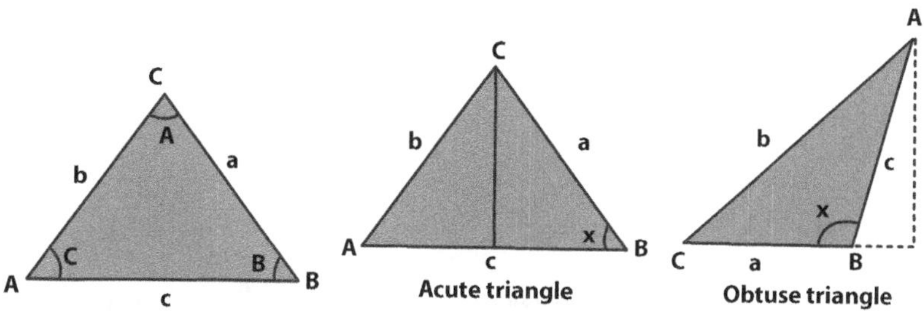

Area of any triangle if the base and height are given.

$$A = \frac{1}{2}\ b \times h$$

Similarly, we have rules to solve and find out the What if we only know any of the two sides and the angle of a given triangle?

This area rule applies to any triangle and states that the area of a triangle is given by half the product of any two sides with the sine of the angle between them.

$$\text{Area} = \frac{1}{2}\ ab \sin C$$

$$\text{Area} = \frac{1}{2}\ bc \sin A$$

$$\text{Area} = \frac{1}{2}\ ac \sin B$$

7.8.3 Ambiguous Case

For the cases (AAS) and (ASA), it is clear to use the sine rule to obtain the other sides of a triangle. Think about the case angle-side-side (ASS), where we are given the measure of just one of the angles in the triangle along with the length of two sides.

In this case, there may be two triangles, one triangle, or no triangle with the given properties. For this reason, it is sometimes called an **ambiguous case**. Thus, we need to examine the possibility of no solution, one or two solutions. Below is a summary of these possible solutions for a triangle.

1. There are three possibilities if A is an acute angle and $a \geq b$.

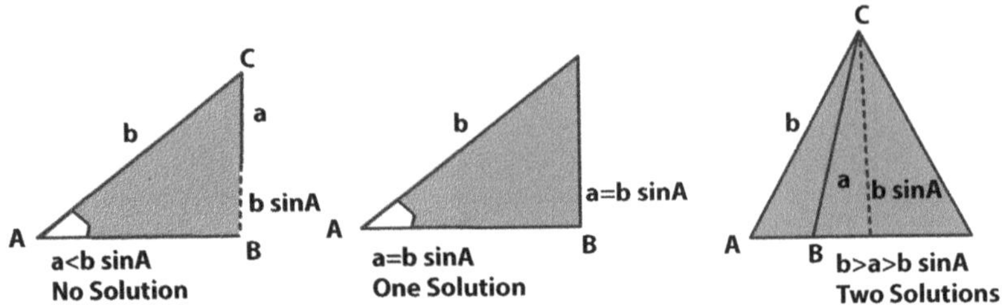

2. If $\angle A$ is an acute and $a \geq b$, then there is exactly one solution.
3. If $\angle A$ is an obtuse or right angle, there are two possibilities.
   a) $a \leq b$ (No solution)
   b) $a > b$ (One solution)

Note: In any triangle, the largest side is opposite to the largest angle.

Case 1: one side and two angles

EXAMPLE 19: Solve the triangle $\triangle ABC$, given that $a = 10$, $A = 41°$ and $C = 75°$
SOLUTION: By subtracting the first two angles from 180°, we can find the third angle.

$\angle A + \angle B + \angle C = 180°$

$\angle B = 180° - \angle A - \angle C$

$= 180° - 41° - 75°$

$\angle B = 64°$

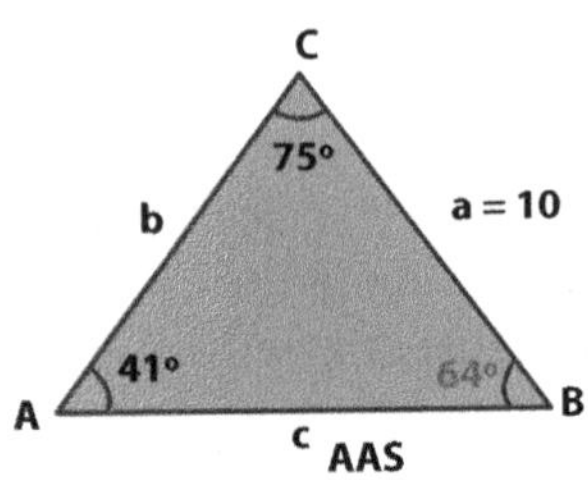

Now, use the sine rule to find the two unknown sides
We have

$$\frac{\sin A}{a} = \frac{\sin B}{b} = \frac{\sin C}{c}$$

Here we can use

$$\frac{sin\ A}{a} = \frac{sin\ C}{c}$$

$$= \frac{sin\ 41°}{10} = \frac{sin\ 75°}{C}$$

$$C = \frac{75°}{sin\ 41°}$$

$$C = 14.7$$

And

$$\frac{sin\ B}{b} = \frac{sin\ A}{a}$$

$$\frac{sin 64°}{b} = \frac{sin 41°}{10}$$

$$b = \frac{10\ sin 64°}{sin 41°}$$

$$b = 13.7$$

Case 2: Two sides and one opposite angle.

EXAMPLE 20: Solve the triangle △ ABC where a=18, ∠A=25°, and b = 30.
SOLUTION: Here, it is given one angle and two sides.

By using the sine rule, we can find out the value of another angle.

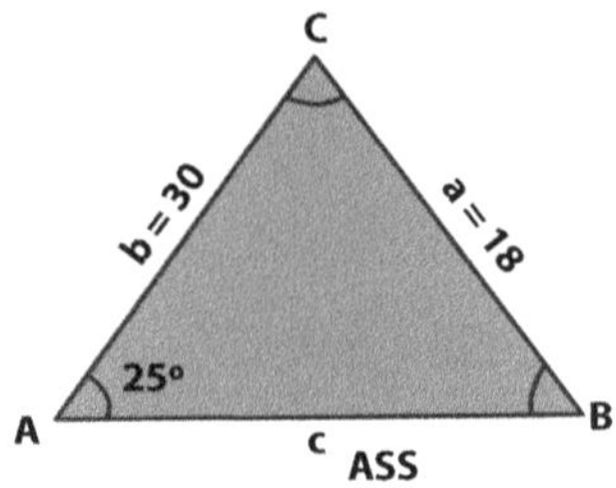

Case 1: $\dfrac{\sin A}{a} = \dfrac{\sin B}{b} = \dfrac{\sin C}{c}$

$$\frac{\sin A}{a} = \frac{\sin B}{b}$$

$$\frac{\sin 25^\circ}{18} = \frac{\sin B}{30}$$

$$\text{sin B} = \frac{25^\circ}{18}$$

$$\text{sin B} = 0.7044$$

$$\text{B} = \sin^{-1}(0.7044)$$

$$\angle \text{B} = 44.8^\circ$$

Now, to find the third angle, we can subtract the sum of the other two angles from the 180°

$$\angle A + \angle B + \angle C = 180^\circ$$

$$25^\circ + 44.8^\circ + \angle C = 180^\circ$$

$$\angle C = 180^\circ - 25^\circ - 44.8^\circ$$

$$\angle C = 110.2^\circ$$

Case 2: As we got $\angle$B = 44.80

But from trig identities, we know that

$$\sin (180^\circ - \boldsymbol{\theta}) = \sin B$$

$$\sin (180^\circ - B) = \sin B$$

$$\sin (180^\circ - 44.8^\circ) = \sin 135.20^\circ = \sin 44.8^\circ$$

So if $\angle B = 135^\circ$

$$C = 180^\circ - A - B$$

$= 180° - 25° - 135.2°$

$C = 19.8°$

Hence if,

∠C = 110.2°	**∠C = 19.8°**
$\frac{c}{\sin C} = \frac{a}{\sin A}$ $\frac{c}{\sin 110.2°} = \frac{18}{\sin 25°}$ $c = \frac{18 \sin 110.2°}{\sin 25°}$ $c = 40$ 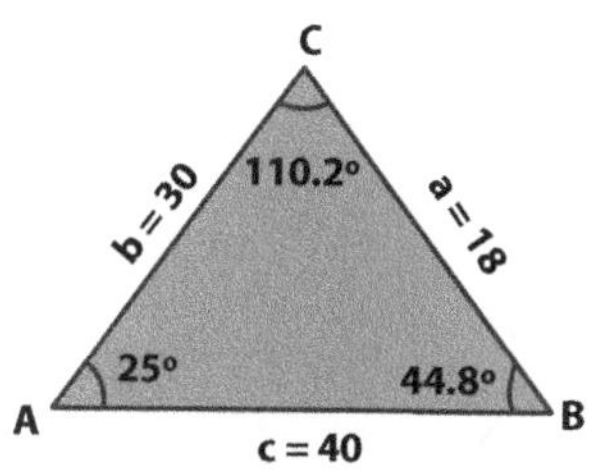 	$\frac{c}{\sin C} = \frac{a}{\sin A}$ $\frac{c}{\sin 19.8°} = \frac{18}{\sin 25°}$ $c = \frac{18 \sin 19.8°}{\sin 25°}$ $c = 14.4$ 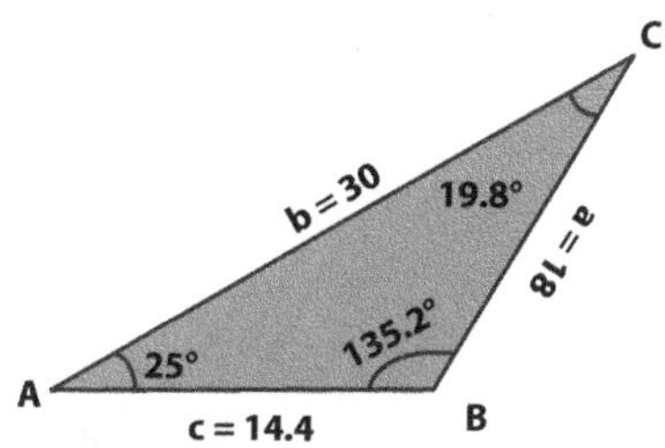

Note: This is an example of an ambiguous case in which there may be more than one solution. In this type of problem, it is also possible for exactly one solution or no solution at all.

Case (2) Another type: Two sides and one opposite angle.

EXAMPLE 21: find the ∠B if in △ABC given that $a=5$, $A = 30°$, and $b = 12$.

SOLUTION: By using the sine rule

$$\frac{\sin B}{b} = \frac{\sin A}{a}$$

$$\frac{\sin B}{12} = \frac{\sin 30^\circ}{5}$$

$$B = \frac{30^\circ}{5}$$

$$\sin B = 1.2$$

$$B = \sin^{-1}(1.2)$$

This is impossible since $|\sin\theta| \leq 1$ for any angle θ.
Thus, there is no solution.

Case 3: Two sides and the angle between them
EXAMPLE 22: solve the triangle $\triangle ABC$ for $\angle A = 30^\circ$, $b = 4$ an $C = 5$
SOLUTION: This is a case of side- Angle – side.

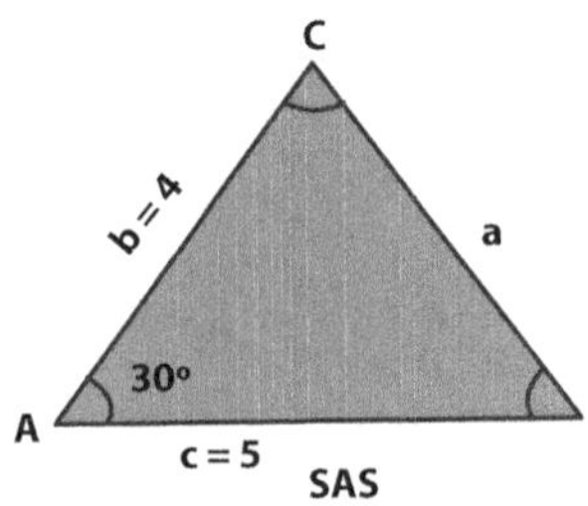

Here we can use the Cosine rule to find the value of a
we have, $a^2 = b^2 + c^2 - 2bc \cos A$
$a^2 = 4^2 + 5^2 - 2(4)(5)\cos 30^\circ$
$a^2 = 6.36$
$a = 2.52$
we can use the cosine rule again to get the value of angle $\angle B$.
We have, $b^2 = c^2 + a^2 - 2ac \cos B$

$$\cos B = \frac{c^2 + a^2 - b^2}{2ca}$$

$$\cos B = \frac{5^2 + (2.52)^2 - 4^2}{2(5)(2.52)}$$

$$\cos B = 0.6091$$

$$B = \cos^{-1}(0.6091)$$

$$B = 52.50°$$

Hence, ∠C = 1800 – ∠A – ∠B
= 180° – 30° – 52.5°
∠C =97.5°

Case 4: Three sides (SSS)
EXAMPLE 23: solve the triangle △ABC if a=2, b=3, and c =4
SOLUTION: The given problem is a case of side-side-side.
Here, we can use the cosine rule to find all angles.

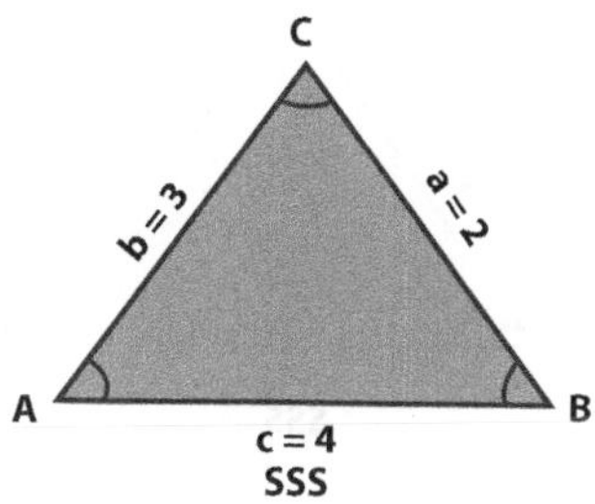

We have

$b^2 = c^2 + a^2 - 2ab \cos B$

$$\cos B = \frac{c^2 + a^2 - b^2}{2ab}$$

$$= \frac{4^2 + 2^2 - 3^2}{2\,(4)(2)}$$

$$= 0.6875$$

$\cos B = 0.6875$
$B = \cos^{-1}(0.6875)$

B = 46.6°

use the cosine rule again to find ∠C.

$$\cos C = \frac{a^2 + b^2 - c^2}{2ab}$$

$$= \frac{2^2 + 3^2 - 4^2}{2\,(2)(3)}$$

$$= \cos C = -\,0.25$$

$$C = 104.5^0$$

Thus, ∠A= 180° - ∠B - ∠C

= 180° – 46.6° – 104.5°

∠A = 28.9°

Case 4: another case of Three sides

EXAMPLE 24: Solve the triangle △ABC if a =2, b=3 and c = 6

SOLUTION: This is a case of side–side–side.

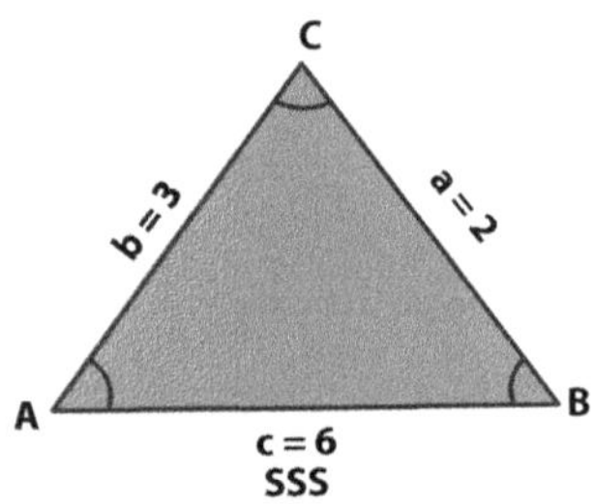

We can directly use the cosine rule to get angles.

$$\text{We have } \cos A = \frac{b^2 + c^2 - a^2}{2ac}$$

$$= \frac{3^2 + 6^2 - 2^2}{2(2)(6)}$$

$$\cos A = 1.139$$

This is impossible since $|\cos A| \leq 1$

Thus, there is no solution.

You can observe that the sum of the length of two sides (2+3=5) is less than the length of the longest side.

As the picture shows, it is impossible in a triangle.

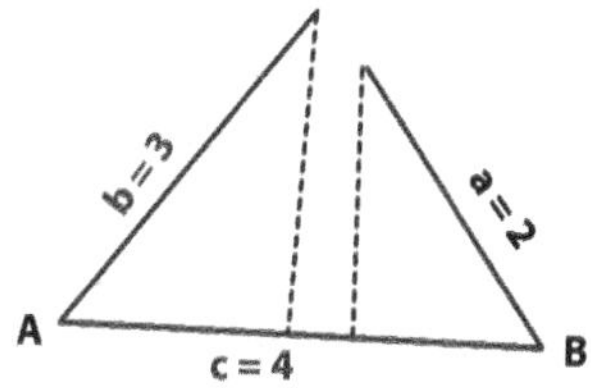

Mixed Level: Exercise 7.6 Practice Questions (Sine and Cosine Rules)

1. Find all the unknown sides and angles of the following triangle
   a. Δ PQR in which ∠ Q = 64°, ∠ R = 24° and r = 3
   b. Δ KLM in which ∠ K = 43°, ∠ M = 50° and m = 1
   c. Δ ABC in which ∠ A = 32.7°, ∠ C = 70.5° and a = 52.3
   d. Δ XYZ in which ∠ X = 56°, ∠Z = 40° and X = 50
2. In Δ ABC, ∠ A = 116°, ∠ C =32°, and AC = 23 m, find the length of the side AB.
3. In Δ RST, ∠ R = 19°, ∠ S=30°, and RT = 120 km. Find the other sides and angels.
4. In Δ KMS, ∠ K = 20°, ∠ M =100°, and S = 23 cm. Find the length of the side m.
5. Solve the following triangles, i.e., find all unknown sides and angles.
   a. Δ ABC, in which ∠ A = 70°, b = 4 and c= 9
   b. Δ XYZ, in which ∠ Y = 112°, x = 2 and y = 3
   c. Δ RST, in which RS = 2, ST = 3 and RT = 5
   d. Δ KLM, in which KL = 5, LM = 10 and KM = 7
   e. Δ JHK, in which ∠ H = 130°, JH = 13 and HK = 8
6. Calculate the third side of the Δ XYZ where
   a. ∠x = 71.4°, y = 3.42 km and z = 4.03 km
   b. x = 103.2 cm ∠Y = 20.80 and z = 44.59 cm
7. Determine the largest angle in
   a. Δ JHK in which JH = 6, HK = 4 and JK = 3
   b. Δ PQR where p = 50, q = 70 and r = 60
8. Find the area of Δ PQR in which
   a. ∠P = 40°, q= 9 and r = 25
   b. ∠Q = 30°, r = 10 and p = 7

c. ∠R = 110°, p = 8 and q = 9
d. PQ = 6 cm PR = 7 cm and ∠R = 28°
e. PR = 52 cm PQ = 29 cm and ∠P = 58.9°

9. Determine the area of a parallelogram in which two adjacent sides are 10cm and 13cm, and the angle between them is 55°.
10. Suppose the area of Δ ABC is 5000 m^2 with a = 150 m and b = 70 m. what are the two possible sizes of ∠ C?
11. Q is a ship at a point 20km due south of another ship. P. the lighthouse R is on the coast such that ∠P = ∠Q = 50°. Find the distance QR.

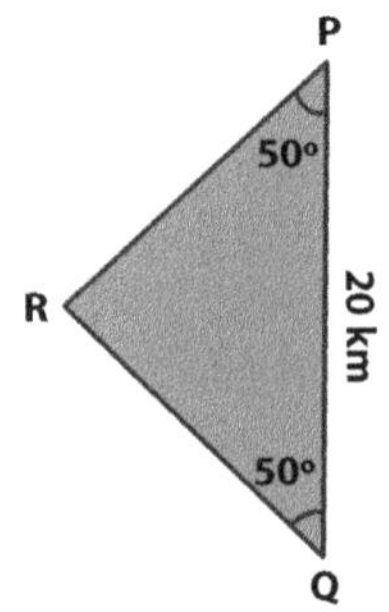

Think and Answer

12. Find the area of the triangle ΔABC given ∠A = 44°, b = 6 units, and c = 9 units

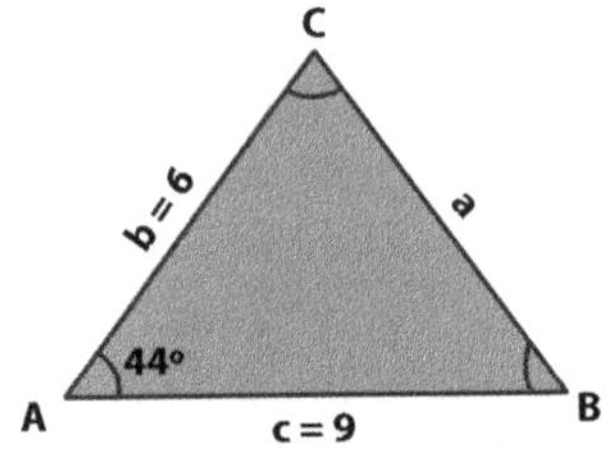

13. Triangular braces are used to support a bridge if the sides of each brace are 75, 52, and 48 meters long. Compute the measure of the angle opposite to the 52 m side.
14. Syed starts his morning walk at point A, reaches two points, B and C, and finally back to A such that ∠A = 60° and

∠B = 45°, AC= 5km in the triangle ABC; find the total distance he covered during his morning walk.

15. Airplane A is flying directly toward the airport, which is 30 miles away. The pilot notices airplane B is 48° to his right. Airplane B is also flying directly toward the airport. The pilot of airplane B calculates that airplane A is 55° to his left, based on the information on how far airplane B is from the airport.

7.9 Steps to Use Calculator for Trigonometry

Find sin 35°

Step 1: Set MODE to Degree.

Step 2: Type entry (Sin 35°) on the Home Screen.

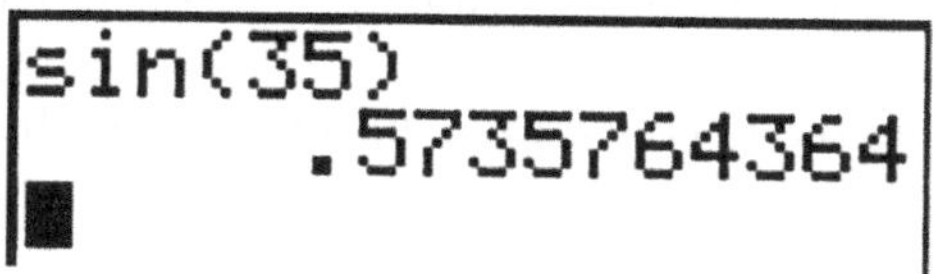

Step 3: You may be asked to round this answer to a specified value. If not requested to round, state the total screen value as the answer. If further calculations are needed, keep this full-screen value and continue working.

Find ∠A to the nearest degree, given cos A = 3/5

Step 1: Set MODE to Degree.

Step 2: On the Home Screen, press Shift + cos to get the $\cos^{-1}$. Type 3/5. Hit ENTER.

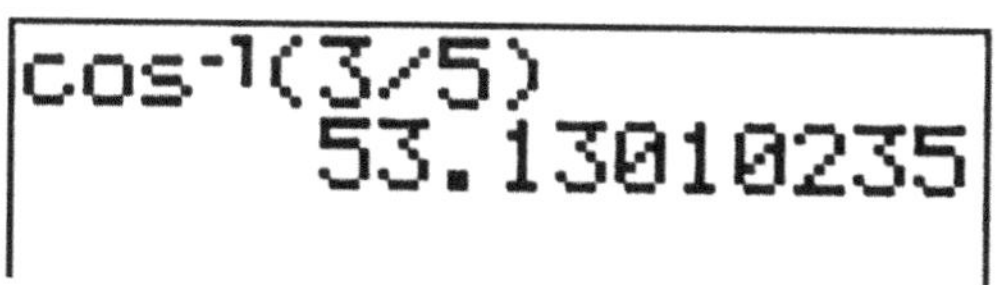

Step 3: Rounding the answer to the nearest degree gives 53°.

Remember, The calculator defaults to the Radian measure.

Converting Degrees to Radians:

Find the radian measure of an angle whose measure is 500°

Step 1: Set MODE to Degree.

Step 2: either Convert by multiplying by $\frac{\pi}{180}$. Or with the mode set to Radian, type 500°. Notice the degree symbol. Hit Enter.

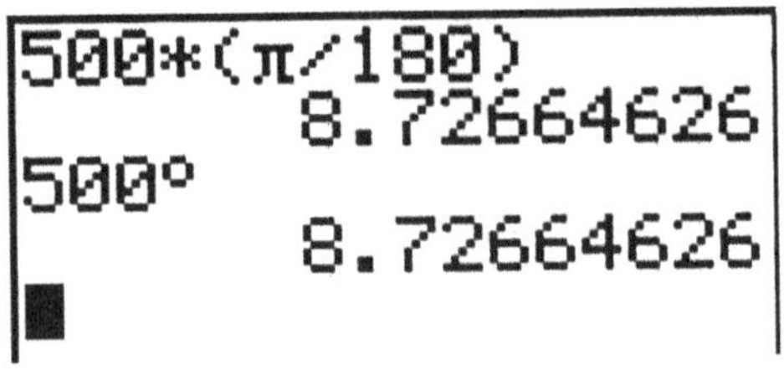

Step 3: Rounding answer to the required decimal places.

Converting Radians to Degrees:

If an angle has a measure $\frac{3\pi}{2}$, find its degree.

Step 1: Set MODE to Degree.
Step 2: either Convert by multiplying by $\frac{180}{\pi}$. Or In Degree mode, the

r symbol can be used to convert radians to degrees. The r is found in 2nd APPS(Angle) #3 r.
Step3: With the mode set to degree, type $(3\pi/2)^r$

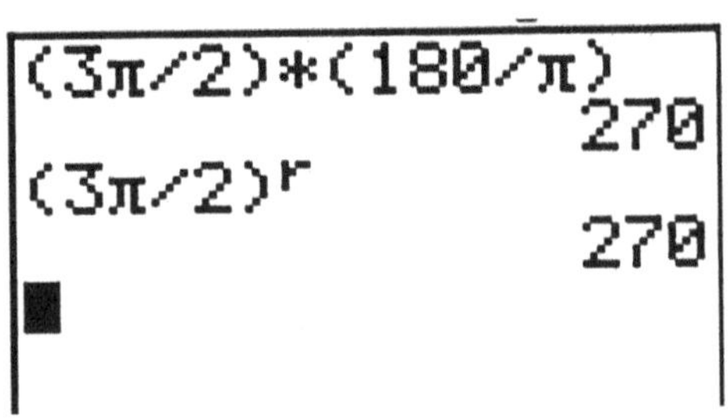

7.10 Check Your Knowledge (MCQ)

1. In a triangle, ABC, AB =28cm, BC = 45cm, and $\angle ABC = 90°$. Determine the length of AC.
   a. 53 cm
   b. 54cm
   c. 43cm
   d. 44cm

2. Determine the value of the sin x of the acute angle x if $\cos x = \frac{2}{\sqrt{10}}$.
   a. $\frac{\sqrt{14}}{2}$
   b. $\frac{\sqrt{14}}{\sqrt{10}}$
   c. $\frac{\sqrt{10}}{\sqrt{14}}$
   d. $\frac{\sqrt{10}}{2}$

3. A 50m building casts a 30m shadow. What is the angle at which the sun hits the building?
   a. 12.4°
   b. 33°
   c. 59°
   d. 95°

4. if tan $x = -3$, determine the value of x in the given interval ($-180° \leq x \leq 180°$).
   a. $\frac{2\pi}{9}, \frac{16\pi}{9}$
   b. 136°, 317°
   c. 93°, 347°
   d. -108°, -72°

5. Determine the arc's length and a sector's area if r = 7cm, θ = 225°.
   a. arc's length=27.49cm, sector's area = 96.21cm^2
   b. arc's length=17.4cm, sector's area = 106.26cm^2
   c. arc's length=56.48cm, sector's area = 45.21cm^2
   d. arc's length=38.87cm, sector's area = 78.98cm^2

6. determine the arc length of a circle if the Area of sector = 440m^2 and θ = 94°.
   a. Area of sector = 504cm^2 and r = 16 cm
   b. Area of sector = 110cm^2 and θ = 35°
   c. Area of sector = 125cm^2 and r = 9 cm
   d. Area of sector = 440m^2 and θ = 94°

7. In Δ DEF, in which d = 4, e = 5, and f = 7, determine the ∠E.
   a. ∠E =44.42°
   b. ∠E =34.05°
   c. ∠E =101.54°
   d. ∠E =56.67°

8. Determine the area of Δ PQR in which QR = 2.5 cm PR = 7.9 cm and ∠R = 125°.
   a. 9.08 unit^2
   b. 7.49 unit^2
   c. 8.09 unit^2
   d. 10.12 unit^2

9. Ali and Hasan cycle apart during a cycling race with an angle of 19° between them. Ali cycles for 5 km, and Hasan cycles for 7 km. How far apart are they?
   a. 3.0 km
   b. 2.8 km

c. 1.5 km
d. 3.8 km

10. Alia and Bayan are standing 25 m apart. Both sites have a helicopter with an angle of elevation o==f 40° and 45°, respectively. How far is the plane from Alia?
a. 18.06m
b. 13.05m
c. 16.75m
d. 17.75m

8
Differential Calculus

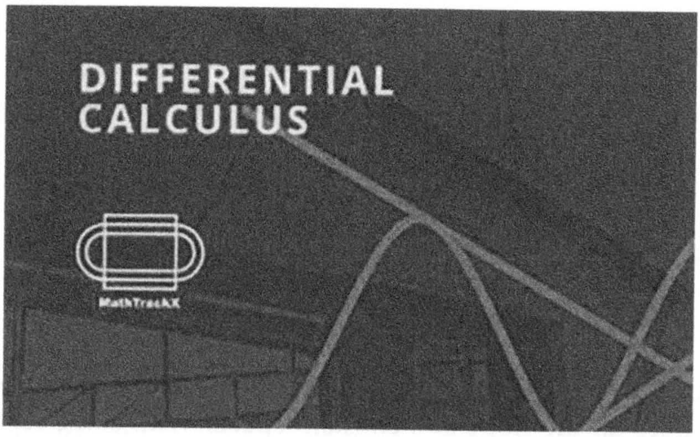

Objectives: This chapter

- *Explains derivatives and their applications.*
- *Introduce some simple rules for finding the gradient of a function.*
- *Apply differentiation to identify the maxima and minima of a function.*

8.1 Introduction

Calculus—"The mathematics of change"

Calculus is one of the most important discoveries in modern mathematics; questions that were difficult to answer by great mathematicians, scientists, and philosophers for more than two thousand years were solved with the help of Calculus. It is just very advanced algebra and geometry. In one sense, it takes the ordinary rules of algebra and geometry and reforms them to be used on more complicated problems.

Ritu Shrivastava, Ramakant Bhardwaj and Satyendra Narayan. Mathematics for Engineers, (325–370)

The use of Calculus is everywhere, mainly in every branch of science and engineering, business, medicine, etc. The fundamental idea of Calculus is to study change by studying "Instantaneous Change," which means changes over a short time.

Let's understand how Calculus is different from simple mathematics. Look at the figure below.

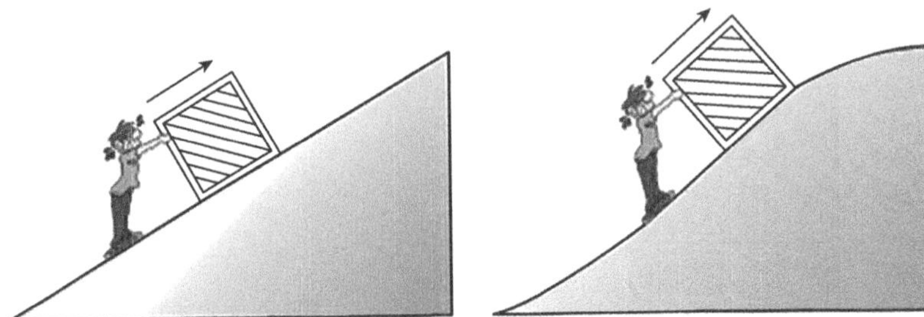

In Figure, Sameer pushes with a constant force for the straight slope, and the crate goes up the slope at a constant speed. With some regular math and simple physics formulas (including algebra and trigonometry), we can calculate the number of calories of energy required to push the crate up the slope. (*Note: The amount of energy expended each second remains the same*)

For the curving slope in Figure, things are constantly changing. The steepness of the slope is constantly changing. And Sameer pushes with a continually evolving force. The vertical the slope, the harder the push, and the amount of energy expended are also changing, not just every second or thousandth of a second, but constantly, from one moment to the next. That is what makes it a calculus problem.

Everything in calculus involves infinity in one way or another because if something is constantly changing, it is changing infinitely from each infinitesimal moment to the next.

Calculus was not discovered all at once. The origins of integral calculus are generally regarded as to the time of the ancient Greeks, circa 200 B.C. Some evidence says that the ancient Egyptians had the idea of Calculus at a much earlier date. In the ancient world, Hellenic mathematician Eudoxus, Archimedes, An Indian mathematician, Bhaskara (1114-1185), 14th-century Indian mathematician Madhava, Jyestadeva, Yuktibhasa, and many other mathematicians developed different results of Calculus. Finally, Issac Newton and Gottfried Leibniz, the two men, are the names behind the discovery of calculus.

The variables x and y were viewed by Leibniz as spanning sequences of endlessly similar values. As disparities between successive values of these sequences, he introduced dx and dy. The tangent was calculated by Newton using the finite velocities "x" and "y." Leibniz is mainly responsible for the notation still used in calculus today.

Reminders

To learn the application of calculus, we need to revise a few definitions.

8.2 Functions

Many mathematical relationships can be considered functions. In mathematics, a function represents the dependence of one quantity upon another. A function means a single letter such as f. If x represents an input to the function, the corresponding output is f(x), read "f of x." The set of all suitable inputs is called the function's **domain**, whereas the set of final output numbers is called the **range** of the function.

8.2.1 Types of Functions

Linear Functions

A function is linear if its slope, or rate of change, is the same at every point. The rate of change of a non-linear function may vary from point to point. Representation of a linear function y = f(x) = mx+ c.

Its graph is a line, where m is the slope or rate of change of y with respect to x. and c is the vertical intercept, or value of y when x is zero.

Exponential Function

We say P is an exponential function of t with the base if $P = P_0$ at, where P_0 is the initial quantity (when t = 0) and a is the factor by which P changes when t increases by 1. We can observe exponential growth when $a < 1$ and exponential decay when $0 < a < 1$.

Composite Functions

Given two functions, f, and g, the composition of f and g is defined by h(x)= f(g(x))

Inverse Functions

If (and only if) a function's graph crosses any horizontal line just once, then it has an inverse.

Logarithm Functions

The Logarithm to base 10 of x, written $\log_{10} x$, is the power of 10 we need to get x. In other words, $\log_{10} x = C$ means $10^c = x$. For the natural Logarithm of x, we often write $\log x = \log_{10} x$. and $\ln x = \log_e x$

Trigonometric Functions

The Sine and Cosine Functions Amplitude, Period, and Phase and The Tangent Functions are known as trigonometric functions.

Power, Polynomials, and Rational Functions

The function of form $f(x) = x^a$, where "a" is a constant, is called a power function. A rational function is a function formed by dividing one polynomial by another: $f(x) = \dfrac{p(x)}{q(x)}$

We begin our calculus study by investigating rates of change for various functions and contexts. Specifically, we see how an average rate of change becomes an instantaneous rate of change through a limiting process. The concept of a limit serves as the foundation for all of calculus; here, we explore limits and their properties and see that the particular type of limit used to find velocities and other rates of change gives rise to the central idea of differential calculus, the derivative. We see how derivatives can be interpreted as rates of change in various situations and the slope of a curve. We learn how the derivative of a function gives information about the original function.

8.3 Limit and Continuity

A limit is described as a number that the function approaches as a variable in an independent function gets closer to a particular value. Examining if a pen can trace a function's graph without removing it from the paper means the function is continuous.

8.3.1 Definition

Limit: Let $f(x)$ be a function with a definition on the interval x that is open and contains the value a. (It is not necessary to define what f(a) is.) If and only if, there exists a number L such that for every > 0, there exists a number L such that $|f(x)-L|<\varepsilon$, whenever $0<|x-a|<\delta$.

This definition is known as ε–δ– or **Cauchy's definition of limit.**

There is also the **Heine definition** of a function's limit, which says that a function f(x) has a limit L at

x = a, if for every sequence $\{x_n\}$, which has a limit at "a". The sequence $f(x_n)$ has a limit of L. The Heine and Cauchy definitions of the limit of a function are equivalent.

8.3.2 One-Sided Limits

Let $\lim\limits_{x \to a-0}$ signify the limit by choosing x values corresponding to the limit as x approaches a. The corresponding limit $\lim\limits_{x \to a-0} f(x)$ is the left-hand limit of f(x) at point x = a.

Similarly, let $\lim\limits_{x \to a+0}$ signify the limit by choosing x values corresponding to the limit as x approaches a. x >a. The corresponding limit $\lim\limits_{x \to a+0} f(x)$ is the right-hand limit of *f(x)* at x = a.

Note that the 2-sided limit $\lim\limits_{x \to a} f(x)$ exists only if the two one-sided limitations exist and are equal.

$\lim\limits_{x \to a-0} f(x) = \lim\limits_{x \to a+0} f(x)$. In this case,

$$\boxed{\lim_{x \to a} f(x) = \lim_{x \to a-0} f(x) = \lim_{x \to a+0} f(x).}$$

8.3.3 Continuity

If, at a given point, there is no break in the graph of a function, the function is **continuous** at that location. A thorough explanation of the continuity of a real function is typically provided in terms of the idea of limits in an introductory calculus course. A function f with variable x is first considered continuous on the real line at the point "a" if the limit of f(x), as x, approaches the point "a," equals the value of f(x) at "a," i.e., f (a). Second, a function is continuous (as a whole) if it is continuous throughout its whole domain.

Continuity can be described mathematically: If all three conditions are met, the function is known as continuous at that specific point.

1. f(a) is defined.
2. $\lim\limits_{x \to a} f(x)$ exists.
3. $\lim\limits_{x \to a+} f(x) = \lim\limits_{x \to a-} f(x) = f(a)$

A function is continuous if you can trace its graph without removing your pen from the page, as was previously stated. Yet, when there is a gap in the middle of a function, it is said to be **discontinuous.**

8.4 Derivatives

A derivative is only a measurement of the rate at which one object changes in relation to another.

8.4.1 Definition

If f(x) is a function of x, then $\lim_{h\to 0}\frac{f(a+h)-f(a)}{h}$ it is defined as the differential coefficient of f(x) at x = a, provided that this limit exists and is finite. The differential coefficient of f(x) with respect to x is generally written as $\frac{d}{dx}$ f(x) or $f'(x)$ or D f(x).

Note: Generally, $\frac{d}{dx}f(x)=\lim_{h\to 0}\frac{f(x+h)-f(x)}{h}$ *is known as the first principle of differential calculus.*

8.4.2 List of Formulae

1. $\frac{d}{dx}C=0$, where C is constant
2. $\frac{d}{dx}x^n=nx^{n-1}$
3. $\frac{d}{dx}Cx=C$, where C is constant
4. $\frac{d}{dx}\log_e x=\frac{1}{x}$
5. $\frac{d}{dx}e^x=e^x$
6. $\frac{d}{dx}a^x=a^x\log_e a$
7. $\frac{d}{dx}x=1$
8. $\frac{d}{dx}\sin x=\cos x$
9. $\frac{d}{dx}\cos x=-\sin x$

10. $\frac{d}{dx}\tan x = \sec^2 x$
11. $\frac{d}{dx}\cot x = -\text{cosec}^2 x$
12. $\frac{d}{dx}\sec x = \sec x \tan x$
13. $\frac{d}{dx}\text{cosec } x = -\text{cosec } x \cot x$
14. $\frac{d}{dx}\log_a x = \frac{1}{x}\log_a e$

8.5 Differentiation Rules

1. Differentiation of Constant Functions

$$\frac{d}{dx}(c) = 0$$

For example: $\frac{d}{dx}(5) = 0$

2. Power Rule

$$\frac{d}{dx}x^n = nx^{n-1} \text{ for all } x$$

For example: Differentiate $\frac{d}{dx}(x^4) = 4x^{4-1}$

3. Differentiation of Constant Multiple Rule

If C is the constant and y is a differentiable function of x, then $\frac{d}{dx}(Cx) = C\frac{d}{dx}(x)$ in particular, if n is any real number, then $\frac{d}{dx}(Cx^n) = Cx^{n-1}$

For example: $\frac{d}{dx}(3x^2) = 3\frac{d}{dx}(x^2) = 3 \times 2x^1 = 6x$

4. Differentiation Sum Rule

If m and n are differential functions of x, then their sum of (m + n) is the differential of every point where m and n are both differentiable at such points.

$$\frac{d}{dx}(m+n) = \frac{d}{dx}m + \frac{d}{dx}n$$

For example: $\frac{d}{dx}(x^3+9x) = \frac{d}{dx}(x^3) + \frac{d}{dx}(9x) = (3x^{(3-1)}) + 9\frac{d}{dx}(x) = 3x^2$
$+9$

5. Differentiation Product Rule

If m and n are differential functions of *x*, then so is their product m and n.

$$\frac{d}{dx}(mn) = \mathrm{m}\frac{d}{dx}n + n\frac{d}{dx}m$$

For example: $\frac{d}{dx}\left[(x+1)(x^2+1)\right]$

Let $\mathrm{m} = x+1,\ \frac{dm}{dx} = 1$

Let $\mathrm{n} = x^2+1,\ \frac{dn}{dx} = 2x$

As we know $\frac{d}{dx}(mn) = \mathrm{m}\frac{d}{dx}n + n\frac{d}{dx}m$

$$\frac{dy}{dx} = (x+1)\frac{d}{dx}(x^2+1) + (x^2+1)\frac{d}{dx}(x+1)$$

$$\frac{dy}{dx} = (x+1)(2x) + (x^2+1)(1)$$

$$\frac{dy}{dx} = (2x^2+2x) + (x^2+1)$$

$$\frac{dy}{dx} = 3x^2+2x+1$$

6. Differentiation Quotient Rule

If m and n are differential functions of *x* if n(*x*) then the Quotient $\frac{m}{n}$ is differential at *x* and

$$\frac{d}{dx}\left(\frac{m}{n}\right) = \frac{n\frac{dm}{dx} - m\frac{dn}{dx}}{n^2}$$

for example: $y = \dfrac{(x-2)(x^2-3x)}{x^3}$

Let, $m = (x-2)(x^2-3x)$

$m = (x^3 - 3x^2 - 2x^2 + 6x) = (x^3 - 5x^2 + 6x)$,

$\dfrac{dm}{dx} = (3x^2 - 10x + 6)$

let, $n = x^3$, $\dfrac{dn}{dx} = 3x^2$

$$\frac{d}{dx}\left(\frac{m}{n}\right) = \frac{n\dfrac{dm}{dx} - m\dfrac{dn}{dx}}{n^2}$$

$$\frac{d}{dx}\left(\frac{m}{n}\right) = \frac{x^3(3x^2-10x+6)-(x^3-5x^2+6x)3x^2}{(x^3)^2}$$

$$\frac{d}{dx}\left(\frac{m}{n}\right) = \frac{(3x^5-10x^4+6x^3)-(3x^5-15x^4+18x^3)}{(x^6)}$$

$$\frac{d}{dx}\left(\frac{m}{n}\right) = \frac{3x^5-10x^4+6x^3-3x^5+15x^4-18x^3)}{(x^6)}$$

$$\frac{d}{dx}\left(\frac{m}{n}\right) = \frac{5x^4-12x^3}{(x^6)}$$

$$\frac{d}{dx}\left(\frac{m}{n}\right) = \frac{x^3(5x-12)}{(x^3)(x^3)}$$

$$\frac{d}{dx}\left(\frac{m}{n}\right) = \frac{5x-12}{x^3}$$

Alternate Method

Expand the numerator and divide it by x^3

$$y = \frac{(x-2)(x^2-3x)}{x^3}$$

$$y = \frac{(x^3-3x^2-2x^2+6x)}{x^3}$$

$$y = \frac{(x^3-5x^2+6x)}{x^3}$$

$$y = \frac{x^3}{x^3} - \frac{5x^2}{x^3} + \frac{6x}{x^3}$$

$$y = 1 - \frac{5}{x} + \frac{6}{x^2}$$

$$y = 1 - 5x^{-1} + 6x^2$$

$$\frac{dy}{dx} = \frac{d}{dx}(1-5x^{-1}+6x^{-2})$$

$$\frac{dy}{dx} = (5x^{-2}-12x^{-3})$$

$$\frac{dy}{dx} = \frac{5}{x^2} - \frac{12}{x^3}$$

$$\frac{dy}{dx} = \frac{5x-12}{x^3}$$

7. Chain Rule

With the help of the chain rule, we can find the derivative of any composite function. If h = f o g is the function such that h(x) = f(g(x)), then

$$\frac{d}{dt}\left[f(g(x))\right] = f'(g(x)) \times g'(x)$$

To understand the above rule, you should recall the concept of a composite function.

A function is composite if you can write it as $f(g(x))$, in other words, its function within a function or function of a function.

Hence the simple idea is.

if $y = f(t), t = g(x)$

then, $\frac{dy}{dx} = \frac{dy}{dt} \times \frac{dt}{dx}$

$\frac{dy}{dt} = (\text{Derivative of outside section}) \times (\text{Derivative of inside section})$

For example: Find the derivative of the following functions. If we find derivative of $y = (3x+1)^2$

Let, $(3x+1) = \mathrm{t}$

$3\mathrm{d}x = \mathrm{dt}, \frac{dt}{dx} = 3$

Hence, $\mathrm{y} = (t)^2, \frac{dy}{dt} = 2\mathrm{t}$

$\frac{dy}{dx} = \frac{dy}{dt} \times \frac{dt}{dx}$

$\frac{dy}{dx} = 2\mathrm{t} \times 3 = 6\mathrm{t} = 6(3x+1)$

EXAMPLE 1: find the derivative of y =2

SOLUTION: $\frac{dy}{dx} = 0$ as we know that the $\frac{d(c)}{dx} = 0$

EXAMPLE 2: find the derivative of $\mathrm{y} = x^1$

SOLUTION: as we know that $\frac{dx^n}{dx} = \mathrm{n}x^{(n-1)}$

hence $\frac{dy}{dx} = 1x^{(1-1)}$

$\frac{dy}{dx} = 1$

$x^0 = 1$

EXAMPLE 3: find the derivative of $y = 2x^1$

SOLUTION: $\frac{d(cx)}{dx} = c\frac{dx}{dx}$

$$\frac{d(2x^1)}{dx} = 2\frac{d}{dx}x = 2 \times 1 = 2$$

EXAMPLE 4: Find the derivatives of $x^{(2/3)}$

SOLUTION: $\frac{dy}{dx}x^{(2/3)} = \frac{2}{3}x^{\left(\frac{2}{3}\right)-1} = \frac{2}{3}x^{\frac{-1}{3}}$

EXAMPLE 5: Find the derivatives of $\frac{1}{x^5}$

SOLUTION: $\frac{d}{dx}\left(\frac{1}{x^5}\right) = \frac{d}{dx}(x^{-5}) = -5x^{-5-1} = -5x^{-6} = \frac{-5}{x^6}$

EXAMPLE 6: Find the derivatives of $x^{\left(-\frac{2}{3}\right)}$

SOLUTION: $\frac{d}{dx}\left(x^{\left(-\frac{2}{3}\right)}\right) = -\frac{2}{3}x^{\left(-\frac{2}{3}\right)-1} = -\frac{2}{3}x^{\frac{-5}{3}} = -\frac{2}{3x^{5/3}}$

EXAMPLE 7: find the derivative of $4x^{\left(-\frac{5}{2}\right)}$

SOLUTION: $\frac{d}{dx}\left(4x^{\left(-\frac{5}{2}\right)}\right) = 4\frac{d}{dx}\left(x^{\left(-\frac{5}{2}\right)}\right)$

$$= 4 \times \left(-\frac{5}{2}\right)x^{\left(-\frac{5}{2}\right)-1} = 4 \times \left(-\frac{5}{2}\right)x^{-\frac{7}{2}}$$

$$= -\frac{20}{2}x^{-\frac{7}{2}} = -10x^{-\frac{7}{2}} = \frac{-10}{x^{\frac{7}{2}}}$$

EXAMPLE 8: find the derivative of $y = x^{\frac{-3}{5}}$

SOLUTION:

$$\frac{dy}{dx} = \frac{d}{dx}\left(x^{\frac{-3}{5}}\right) = \frac{-3}{5}(x)^{\left(\frac{-3}{5}-1\right)}$$

$$= \frac{-3}{5}(x)^{\left(\frac{-8}{5}\right)}$$

$$= \frac{-3}{5x^{\left(\frac{8}{5}\right)}}$$

EXAMPLE 9: find the derivative of $y = 3x^5$

SOLUTION: $\frac{dy}{dx} = 3\frac{dx^5}{dx}$

$$\frac{dy}{dx} = 3 \times 5x^{5-1} = 15x^{5-1} = 15x^4$$

EXAMPLE 10: find the derivative of $y = \frac{5}{x}$

SOLUTION: $y = 5x^{-1}$,

$$\frac{d}{dx}(5x^{-1}) = 5(-1)x^{(-1-1)}$$

$$= 5(-1)x^{-2} = \frac{(-5)}{x^2}$$

EXAMPLE 11: find the derivative of $y = \sqrt{x}$

SOLUTION: $\frac{d(\sqrt{x})}{dx} = \frac{d(x^{1/2})}{dx}$

$$= \frac{1}{2}x^{\left(\frac{1}{2}\right)-1} = \frac{1}{2}x^{\left(-\frac{1}{2}\right)}$$

$$= \frac{1}{2}\frac{1}{x^{\left(\frac{1}{2}\right)}} = \frac{1}{2\sqrt{x}}$$

EXAMPLE 12: find the derivative of $y = 3x^3 - 4x^2 + 3x - 1$

SOLUTION: $\frac{d}{dx}\left(3x^3 - 4x^2 + 3x - 1\right)$

$$= \frac{d}{dx}\left(3x^3\right) - \frac{d}{dx}\left(4x^2\right) + \frac{d}{dx}(3x) - \frac{d}{dx}(1)$$

$$= 3\frac{d}{dx}\left(x^3\right) - 4\frac{d}{dx}\left(x^2\right) + 3\frac{d}{dx}(x) - \frac{d}{dx}(1)$$

$$= 3\left(3x^2\right) - 4\left(2x^1\right) + 3(1) - (0)$$

$$= 9x^2 - 8x^1 + 3$$

EXAMPLE 13: find the derivative of $y = x^4 + \frac{2}{3}x^3 - 7x^2 + 5x + 1$

SOLUTION: $\frac{dy}{dx} = \frac{d}{dx}\left(x^4\right) + \frac{d}{dx}\left(\frac{2}{3}x^3\right) - \frac{d}{dx}\left(7x^2\right) + \frac{d}{dx}(5x) + \frac{d}{dx}(1)$

$$\frac{dy}{dx} = \frac{d}{dx}\left(x^4\right) + \frac{2}{3}\frac{d}{dx}\left(x^3\right) - 7\frac{d}{dx}\left(x^2\right) + 5\frac{d}{dx}(x) + \frac{d}{dx}(1)$$

$$\frac{dy}{dx} = \left(4x^{4-1}\right) + \frac{2}{3} \times 3\left(x^{3-1}\right) - 7 \times 2\left(x^{2-1}\right) + 5 \times 1\left(x^{1-1}\right) + 0$$

$$\frac{dy}{dx} = \left(4x^3\right) + \frac{2}{3} \times 3\left(x^{3-1}\right) - 7 \times 2\left(x^{2-1}\right) + 5 \times 1\left(x^{1-1}\right) + 0$$

$$\frac{dy}{dx} = 4x^3 + 2x^2 - 14x + 5$$

EXAMPLE 14: find the derivative of $y=(3x-1)(4x+5)$

SOLUTION: By, Product rule

$$\frac{d}{dx}(mn)=\mathrm{m}\frac{d}{dx}n+n\frac{d}{dx}m$$

$$\frac{dy}{dx}=\frac{d}{dx}(3x-1)(4x+5)=(3x-1)\frac{d}{dx}(4x+5)+(4x+5)\frac{d}{dx}(3x-1)$$

$$=(3x-1)(4)+(4x+5)(3)$$

$$=(12x-4)+(12x+5)$$

$$=24x+11$$

EXAMPLE 15: Find the derivatives of $f(x)=y=(x^2+5x-6)^9$

SOLUTION: $let,\ t=x^2+5x-6,\ \frac{dt}{dx}=2x+5$

$$\mathrm{y}=\mathrm{t}^9,\ \frac{dy}{dt}=9\mathrm{t}^8$$

$$\frac{dy}{dx}=\frac{dy}{dt}\times\frac{dt}{dx}$$

$$\frac{dy}{dx}=9(t)^8\times(2x+5)$$

$$\frac{dy}{dx}=9(x^2+5x-6)^8\times(2x+5)$$

EXAMPLE 16: find the derivative of $y = (2x+3)(5x^2 - 4x)$

SOLUTION: $\frac{dy}{dx} = \frac{d}{dx}\left[(2x+3)(5x^2-4x)\right]$

Let m $= (2x+3)$, $\frac{dm}{dx} = 2$

n $= (5x^2 - 4x)$, $\frac{d\text{n}}{dx} = 10x - 4$

$$\frac{d}{dx}(mn) = m\frac{dn}{dx} + n\frac{dm}{dx}$$

$$\frac{d}{dx}(mn) = (2x+3)(10x-4) + (5x^2-4x)\times 2$$

$$\frac{d}{dx}(mn) = (20x^2 - 8x + 30x - 12) + (10x^2 - 8x) = (30x^2 + 14x - 12)$$

EXAMPLE 17: find the derivative of $y = (x^2+1)(x+5+\frac{1}{x})$

SOLUTION: Let m $= (x^2+1)$, $\frac{dm}{dx} = 2x$

n $= \left(x+5+\frac{1}{x}\right)$, $\frac{d\text{n}}{dx} = 1 - \frac{1}{x^2}$

$$\frac{d}{dx}(mn) = m\frac{dn}{dx} + n\frac{dm}{dx}$$

$$\frac{d}{dx} = (x^2+1)\left(1-\frac{1}{x^2}\right) + \left(x+5+\frac{1}{x}\right)2x$$

$$= (x^2 - 1 + 1 - \frac{1}{x^2}) + (2x^2 + 10x + 2) = 3x^2 + 10x + 2 - \frac{1}{x^2}$$

EXAMPLE 18: find the derivative of $y = \frac{(x^2+1)}{x-1}$

SOLUTION: By quotient rule

$$\frac{d}{dx}\left(\frac{m}{n}\right) = \frac{n\frac{dm}{dx} - m\frac{dn}{dx}}{n^2}$$

$$\frac{dy}{dx} = \frac{d}{dx}\left[\frac{(x^2+1)}{(x-1)}\right] = \frac{(x-1)\frac{d}{dx}(x^2+1) - (x^2+1)\frac{d}{dx}(x-1)}{(x-1)^2}$$

$$\frac{dy}{dx} = \frac{(x-1)(2x) - (x^2+1)(1)}{(x-1)^2}$$

$$\frac{dy}{dx} = \frac{(2x^2-2x) - (x^2+1)}{(x-1)^2}$$

$$\frac{dy}{dx} = \frac{2x^2-2x-x^2-1}{(x-1)^2}$$

$$\frac{dy}{dx} = \frac{x^2-2x-1}{(x-1)^2}$$

EXAMPLE 19: find the derivative of $y = \frac{2x+7}{3x-4}$

SOLUTION: Let $m = 2x+7$, $\frac{dm}{dx} = 2$

and $n = 3x-4$, $\frac{dn}{dx} = 3$

Use the quotient rule

$$\frac{d}{dx}\left(\frac{m}{n}\right)=\frac{n\frac{dm}{dx}-m\frac{dn}{dx}}{n^2}$$

$$\frac{d}{dx}\left(\frac{2x+7}{3x-4}\right)=\frac{(3x-4)(2)-(2x+7)\times 3}{(3x-4)^2}$$

$$\frac{d}{dx}\left(\frac{2x+7}{3x-4}\right)=\frac{(6x-8)-6x-21)}{(3x-4)^2}=\frac{-29}{(3x-4)^2}$$

EXAMPLE 20: find the derivative of $f(t)=\frac{t^2-4}{(t^2+4t+4)}$

SOLUTION: The derivative of the above function can be found using the quotient rule. Let's simplify the rule before applying it.

$$f(t)=\frac{t^2-4}{(t^2+4t+4)}$$

$$f(t)=\frac{t^2-2^2}{(t^2+4t+4)}$$

$$f(t)=\frac{(t+2)(t-2)}{(t+2)^2}$$

$$f(t)=\frac{\cancel{(t+2)}(t-2)}{\cancel{(t+2)}(t+2)}$$

$$f(t)=\frac{(t-2)}{(t+2)}$$

$$\text{Let, m}=t-2,\ \frac{dm}{dt}=1$$

$$\text{Let, n}=\text{t}+2,\ \frac{an}{dt}=1$$

$$\frac{d}{dt}f(t)=\frac{(t+2)(1)-(t-2)(1)}{(t+2)^2}$$

$$\frac{d}{dt}f(t)=\frac{t+2-t+2}{(t+2)^2}=\frac{4}{(t+2)^2}$$

Mixed Level: Exercise 8.1 Practice Questions (Derivatives of Algebraic Functions)

1. Find the derivative of
   a. $y = 3x^5$
   b. $y = 3x^5 - 2x^4 + x^3 - x^2 + 1$
   c. $y = \frac{x^5 - 1}{x^2 + 1}$
   d. $y = \sqrt{2x + 5}$
   e. $y = \frac{-1}{4}x^{1/2}$
   f. $y = \frac{1}{\sqrt{x}}$
   g. $y = \frac{-x^2}{x^2 + 1}$
   h. $y = (x^3 - x^2 + 1)(x + 1)$
   i. $y = \frac{x^3 + 2x^2 - 1}{x + 7}$

2. Find the derivative of
   a. $y = \frac{4}{3x^{3/2}}$
   b. $y = -3x^2 - 7$
   c. $y = \frac{x^3 - 5}{3}$
   d. $y = 2x^4 + x^3 - x^2 + 4$
   e. $y = (x^3 + 3x - 2)^4$
   f. $y = \frac{3}{x^3} + \frac{2}{x^2}$
   g. $y = \sqrt{x^2 - 2x + 3}$
   h. $y = 2t^2 - 10t$
   i. $y = t^{-4} - 9t^{-3} + 8t^{-2} - 10$
   j. $y = \frac{4}{t} - \frac{1}{6t^2} + \frac{8}{t^5}$
   k. $y = (t - 4)(2t^2 - t^3)$
   l. $y = \frac{y^5 - 5y^3 + 2}{y^2}$
   m. $y = \frac{3}{\sqrt{m^3}} + \frac{8}{5m^4} - \frac{1}{3m^9}$

8.6 Derivatives of Logarithmic and Exponential Function

8.6.1 Laws of Exponents

1. $a^0 = 1$
2. $a^m \times a^n = a^{m+n}$
3. $\frac{a^m}{a^n} = a^{m-n}$
4. $\frac{1}{a^n} = a^{-n}$
5. $(a^m)^n = a^{mn}$
6. $\sqrt{a} = a^{\frac{1}{2}}$
7. $\sqrt[n]{(a)^m} = a^{\frac{m}{n}}$
8. $(a.b)^m = a^m a^n$
9. $\sqrt{\frac{a}{b}} = \frac{\sqrt{a}}{\sqrt{b}}$
10. $\sqrt{ab} = \sqrt{a}\sqrt{b}$

8.6.2 Rules of Logarithm

1. $\log_a b = 1$
2. $\log_a 1 = 0$
3. $\log_a m + \log_a n = \log_a mn$
4. $\log_a m - \log_a n = \log_a \frac{m}{n}$
5. $n \log_b m = \log_b m^n$

8.6.3 Rules for Finding the Derivative of Log and Exp

$$y = \log x, \ \frac{dy}{dx} = \frac{1}{x}$$